*Dedicated to our wives Azadeh Alimdad
and Anjali Mago for their patience
and support*

Acknowledgments

We are extremely grateful to the IRMACS Center at Simon Fraser University, Burnaby, Canada for the administrative, and technical support that greatly facilitated this research.

We also wish to thank the reviewers for their constructive comments on the chapters.

Laurens Bakker	KPMG, Netherlands
Nitin Bhatia	DAV College, Jalandhar, India
Gabriel Burstein	City University of New York, New York, USA
Thomas Couronné	Orange Labs/France Télécom R&D, France
Suzana Dragićević	Simon Fraser University, Burnaby, Canada
Richard Frank	Simon Fraser University, Burnaby, Canada
Philippe Giabbanelli	Simon Fraser University, Burnaby, Canada
Piper Jackson	Simon Fraser University, Burnaby, Canada
Manpreet Jammu	Microsoft, Washington, USA
Mamta Khosla	Dr. B. R. Ambedkar National Institute of Technology, Jalandhar, India
Vijay K Mago	Simon Fraser University, Burnaby, Canada
Bimal Kumar Mishra	Birla Institute of Technology, Ranchi, India
Susan Mniszewski	Los Alamos National Laboratory, USA
Elpiniki I. Papageorgiou	Technological Educational Institute of Lamia, Greece
Laurent Tambayong	University of California, Irvine, USA
Pietro Terna	University of Torino, Italy
Katie Wuschke	Simon Fraser University, Burnaby, Canada

Contents

Chapter 1
Introducing Theories and Simulations of Complex Social Systems

Vahid Dabbaghian and Vijay Kumar Mago

1 Introduction

Complex networks exist throughout the domains of nature and society: swarms, circulatory systems, roads, power grids. These networks enable the efficient distribution of resources, resulting in greater and more impressive activity. Social systems are networks that go one step further: they are not only there for the distribution of resources, but also to act as a medium for the interaction between numerous intelligent entities. Thus they combine efficient resource use with intense productivity, in the sense that interactions between these entities produce numerous effects on the system, for better or for worse [1]. We live our lives in a nexus of numerous social systems: family, friends, organizations, nations. We benefit from the fruits of their power, including energy, learning, wealth, and culture. We also struggle with the crises they generate, such as crime, war, pollution, and illness. Our motivation for studying these systems is clear: they are the fabric upon which our lives are woven.

Research into social systems is challenging due to their complex nature. Traditional methods of analysis are often difficult to apply effectively. This can be due to a lack of appropriate data, or too much uncertainty. It can also be the result of problems which are not yet understood well enough in the general sense so that they can be classified, and an appropriate solution quickly identified. Simulation is one tool that deals well with these challenges, fits in well with the deductive process, and is useful for testing theory [4]. This field is still relatively new, and much of the work is necessarily innovative, although it builds upon a rich and varied foundation [6]. There are a number of existing modelling paradigms being applied to complex social systems

V. Dabbaghian (✉) · V. K. Mago
The Modelling of Complex Social Systems, The IRMACS Centre,
Simon Fraser University, Burnaby, Canada
e-mail: vdabbagh@sfu.ca

V. K. Mago
e-mail: vmago@sfu.ca

V. Dabbaghian and V. K. Mago (eds.), *Theories and Simulations of Complex
Social Systems*, Intelligent Systems Reference Library 52,
DOI: 10.1007/978-3-642-39149-1_1, © Springer-Verlag Berlin Heidelberg 2014

research [2, 3, 7–9]. Additionally, new methods and measures are being devised through the process of conducting research. It is vital to point out that this novelty is not a reason for hesitation in following this line of research: the problems under consideration here are of significant concern [5]. The intensification of activity in modern life means that the consequences for mistakes or inaction in public policy are grave. This is true for political, financial, justice, and public health decisions, among many other fields. We present in this volume a selection of research that seeks to address the challenges being faced in these fields, both in terms of finding solutions and the best ways to pursue this kind of research.

2 New Contributions

New directions in science require new ways of thinking about the practice of science. This volume begins with Chap. Software Solutions for Computational Modelling in the Social Sciences with discussion of the challenges of creating the software which underlies modelling research of social systems. A general approach for the development process is presented, along with solutions to common issues. This approach takes advantage of the dynamic nature of software so that it can be effectively used as a tool of discovery and communication. In Chap. Modelling Epistemic Systems, Martins seeks to model the process by which a community of scientists produces scientific knowledge. He employs the ideas considered by precursors in this field and builds upon them by using a computational model to investigate different ideas through experimentation. This extension of traditional approaches through the application of computational experimentation demonstrates some of the new ways in which computational research is expanding the social sciences.

In Chap. Modeling Human Behavior in Space and Time Using Mobile Phone Data, Couronne discusses the properties of mobile phone tracking data. With the ubiquity of mobile phone use in both developed and developing countries, this kind of data has great potential for providing insight into many aspects of human activity. Couronn outlines some of the issues of using this data and describes useful metrics for this purpose. Chapter Change Detection in Dynamic Political Networks: The Case of Sudan derives its data from a more traditional source, newspapers, to consider a recent area of instability. Tambayong and Carley analyse political networks in Sudan through the application of an algorithm to detect social network change. In this way, they are able to pinpoint network behaviour that matches historical activity. In these works, mathematical modelling provides a way to analyze data and encapsulate salient features lying therein.

Complex social systems are found throughout the gamut of large-scale human interaction, including activities of great concern to social policy decision making. The next two chapters consider criminal justice as their topic. In Chap. High-Level Simulation Model of a Criminal Justice System, Dabbaghian et al. present a system dynamics model of the criminal justice system in British Columbia, Canada. This high-level model incorporates the numerous component systems necessary for the wheels of justice to turn, and is designed to give stakeholders better understanding of how the system works as a whole, and how it will behave in certain situations. In Chap. Celerity in the Courts:The Application of Fuzzy Logic to Model Case Complexity Criminal Justice Systems, Reid and Frank propose a methodology for increasing the efficiency of the criminal justice by identifying complex cases early on in the process. Their proposal adopts the expressive capabilities of fuzzy logic to mediate uncertain and ambiguous factors involved in a mathematical manner.

A different kind of interaction that is also of significant concern to policy makers is the widespread transmission of disease. Mniszewski et al. tackle the issue of face-mask use (to prevent the transmission of disease) in Chap. Understanding the Impact of Face Mask Usage through Epidemic Simulation of Large Social Networks. They adopt agent-based simulation to consider face-mask use within the context of an influenza outbreak. Their research underlines the need for a warning that masks alone are not sufficient to fight disease transmission: an integrated approach including education and monitoring is more promising. Chapter e-Epidemic Models on the Attack and Defense of Malicious Objects in Networks (Mishra and Haldar) consider not biological disease but technological: the proliferation of attacks on computer systems. They adopt modelling techniques from the health sciences designed for epidemiology to the aim of developing comprehensive models and methods for dealing with this problem. While these chapters consider fundamentally different kinds of disease, modelling is useful in both cases to capture the essential nature of the phenomena under study.

This volume also includes works which highlight the capabilities of contemporary modelling techniques. Chapter Modelling the Joint Effect of Social Determinants and Peers on Obesity Among Canadian Adults (Giabbanelli et al.) produces results that suggest that network qualities affect the way in which obesity prevention is transferred through social connections. This research is based on a well-founded Fuzzy Cognitive Map (FCM) modelling obesity trends, processed in parallel across a simulated social network. In Chap. Youth Gang Formation: Basic Instinct or Something Else?, Morden et al. apply FCMs to the age-old problem of criminal youth gangs. Here also, FCMs appeal because of their ability to capture interconnected domain knowledge in a theoretical object with useful mathematical properties.

Agent-based modelling (ABM) focuses on the activities and interactions of individual humans within a larger environment. The close ontological match between model and our perception of the phenomena has made ABM a favoured choice for research into complex social systems. Malleson et al. (Chap. Optimising an Agent-Based Model to Explore the Behaviour of Simulated Burglars) combines this technique with a genetic algorithm to model the target selection behaviour of burglars. Problems in criminology are notably difficult to obtain complete data for due to the nature of the field: criminals are unsurprisingly reticent to reveal all the details of their activities. As can be seen here, simulation provides a venue for combining expert knowledge and logic in order to come to better conclusions.

3 Conclusion

The research contained in this volume demonstrates how modelling is being used in the social sciences and public health to formalize knowledge, develop theories and perform experimentation via simulation. We see a different approaches and techniques being applied to unrelated problem domains. This new wave of science has been enabled by advances in computing technology, but its ongoing impact on a variety of sciences and related disciplines is manifold and pervasive. Successes in both experimental results and methodology feed into each other and further reinforce a foundation of knowledge and practice upon which to base evolving vectors of research. The selection of works presented here each seek to meaningfully advance science in their own way, and it is our sincere hope that you will agree that they are valuable and interesting contributions to this dynamic and growing area of inquiry.

References

1. Bettencourt, L.M., Lobo, J., Helbing, D., Kühnert, C., West, G.B.: Growth, innovation, scaling, and the pace of life in cities. Proc. Nat. Acad. Sci. U.S.A. **104**(17), 7301–7306 (2007)
2. Dabbaghian, V., Jackson, P., Spicer, V., Wuschke, K.: A cellular automata model on residential migration in response to neighborhood social dynamics. Math. Comput. Model. **52**(9), 1752–1762 (2010)
3. Dabbaghian, V., Spicer, V., Singh, S.K., Borwein, P., Brantingham, P.: The social impact in a high-risk community: a cellular automata model. Int. J. Comput. Sci. **2**(3), 238–246 (2011)
4. Epstein, J.M.: Agent-based computational models and generative social science. In: Generative Social Science: Studies in Agent-Based Computational Modeling, pp. 4–46 (1999)
5. Funtowicz, S.O., Ravetz, J.R.: Science for the post-normal age. Futures **25**(7), 739–755 (1993)
6. Gilbert, G.N.: Computational Social Science. Sage, London (2010)
7. Simon Fraser University. Complex Systems Modelling Group.: Modelling in Healthcare. American Mathematical Society, Providence, RI, USA (2010)
8. Mago, V.K., Bakker, L., Papageorgiou, E.I., Alimadad, A., Borwein, P., Dabbaghian, V.: Fuzzy cognitive maps and cellular automata: an evolutionary approach for social systems modelling. Appl. Soft. Comput. **12**(12), 3771–3784 (2012)
9. Mago, V.K., Frank, R., Reid, A., Dabbaghian, V. (2013): The strongest does not attract all but it does attract the most-evaluating the criminal attractiveness of shopping malls using fuzzy logic. Expert Syst. doi:10.1111/exsy.12015, http://onlinelibrary.wiley.com/doi/10.1111/exsy.12015/full

Chapter 2
Software Solutions for Computational Modelling in the Social Sciences

Piper J. Jackson

Abstract Social science is critical to decision making at the policy level. New research in the social sciences focuses on using and innovating new computational methods. However, as a relatively new science, computational research into the social sciences faces significant challenges, both in terms of methodology and acceptance. Important roles software can take in the process of constructing computational models of social science phenomena are discussed. An approach is presented that frames software within this kind of research, aiming at involving software at an early stage in the modelling process. It includes a software framework that seeks to address the issues of computational social science: iterative knowledge development; verification and validation of models; and communication of results to peers and stakeholders.

1 Introduction

With the emergence of cheap and powerful computation, social scientists have started to explore the potential of applying computational to their research topics [1]. Hummon and Fararo claim that the traditional two component model of science—theory and empirical research—needs to be expanded to include computation as its third component. Simulation can be thought of as the interaction of theory and computation components. High-level programming code, with the capacity to clearly represent both entities (data structures) and behaviour (algorithms), provides a conceptual

P. J. Jackson (✉)
Modelling of Complex Social Systems (MoCSSy) Program, Interdisciplinary Research in the Mathematical and Computational Sciences (IRMACS) Centre, Simon Fraser University, 8888 University Drive, Burnaby, BC V5A 1S6, Canada
e-mail: pjj@sfu.ca

P. J. Jackson
Software Technology Lab, School of Computing Science, Simon Fraser University, 8888 University Drive, Burnaby, BC V5A 1S6, Canada

V. Dabbaghian and V. K. Mago (eds.), *Theories and Simulations of Complex Social Systems*, Intelligent Systems Reference Library 52,
DOI: 10.1007/978-3-642-39149-1_2, © Springer-Verlag Berlin Heidelberg 2014

link between the underlying mathematics of the simulation model and researchers' understanding. Being able to test theories using mathematics promises insight formerly limited to the physical sciences, but it is important to remember that social structures are always at least partly interpretive in nature, since they are constructed and maintained by human activity. Gilbert claims that this is not problematic, and indeed may even be more faithful a representation than simulation of physical phenomena: a translation from a (social) construct to a (computational) construct is likely to be less problematic than translating something in the real world into a constructed computer representation [2].

For the social sciences, applying computational techniques helps in overcoming some of the core limitations of studying social phenomena. Social scientists have always been limited by the inextricability of the subject of their research from its environment. Hence, it is difficult to study different factors influencing a phenomena in isolation. Safety and ethical issues can be an obstacle to innovation, e.g., for criminologists, it is very difficult to get first-hand evidence of crimes while they are being perpetrated—an observer would most likely be legally required to try to prevent the crime rather than letting it take place. Developing response strategies to unpredictable and dangerous situations is difficult to do in the field, since such situations are unpredictable and by their nature very difficult to control. Computational methods allow us to circumvent these problems by generating scenarios inside a virtual environment. In particular, modelling and simulation allow us to dynamically and interactively explore our ideas and existing knowledge.

Computational thinking about social phenomena, however, means thinking in terms of multiple layers of abstraction [3], which facilitates a systematic study of the phenomena by adjusting the level of detail given to the various factors under study. Computer models of social systems simulate dynamic aspects of individual behaviour to study characteristic properties and dominant behavioural patterns of societies as a basis for reasoning about real-world phenomena. This way, one can perform experiments as a mental exercise or by means of computer simulation, analyzing possible outcomes where it is difficult, if not impossible, to observe such outcomes in real life.

However, when we speak of computational methods, it is easy to gloss over the software that is the means to our research goals. We have an intuitive model of what a software program should be, and it is easy to apply this in a superficial sense to the problem of building a social system simulation. Yet the characteristics and functionality of that software are of great importance to our work, as are the roles which we want it to play in our research endeavours. The aspects that make software successful in an interdisciplinary computational social science research project will be outlined here. From these roles, a framework has been developed that is presented here. This framework aims at circumventing problems commonly faced, and increasing the productivity and advancement of groups working on these kind of simulation research projects.

The chapter is organized as follows. Related work is described in Sect. 2. Issues related to the software development process are discussed in Sect. 3. Section 4 describes the main pillars of a social science computational modelling project, as well as the characteristics such an endeavour should possess. Section 5 is a presentation of the framework, followed by some final words in Sect. 6.

2 Related Work

Discussion and guidelines for how social simulations can be developed using the input of multiple stakeholders is presented in [4]. Importantly, human factors, such as maintaining stakeholder motivation, and technical factors, such as the importance of a formal computing models, are emphasized in their approach.

A particularly complete methodology for prototyping and developing agent-based systems is presented in [5]. The methodology described, ELDAMeth, includes an overall development lifecycle model, automated code generation based on high-level models, and testing through simulation of the target system.

There are numerous multi-agent modelling toolkits available, which include programming libraries and/or languages that facilitate the production of simulation software. A common and important component are methods for visualizing models graphically, reducing the need for toolkit users to have specialist technical knowledge in that kind of programming. Two toolkits are of particular note. Repast has a wide user base and allows for the creation of models using ReLogo (a dialect designed for Repast), Java, and flowcharts [6]. MASON is another toolkit with which a variety of model types can be developed, and is notable in that both the simulation itself and its data visualization can be run concurrently [7].

More generally, Nigel Gilbert and his many colleagues have been central in promoting the use of simulation models in social science research. One example is [8], in which the motivation and general methodology for using such models for social simulation is presented. Other work in this field has suggested common approaches and vocabulary to describe agent-based social simulations [9].

3 Software Development

Software development methodologies enable systematic production of software. Agile methodologies have found widespread adoption in industry due to higher rates of success while using fewer resources. Here success is in terms of getting a project done on time and on budget. The category of agile methodologies includes Extreme Programming (XP), the Rational Unified Process (RUP) and Scrum, among others. Agile methods are diverse in practice, but share a number of important characteristics. First of all, they emphasize the role of people in the development process. Thus, how people interact and relate to each other is considered as an important factor in the production of software. This includes clients as well as the people working on the software. They also emphasize flexibility in the face of change and obstacles, which is appropriate since this is a strength of software itself. A particularly important characteristic of agile methods is their subscription to an iterative mode of production. Instead of following an initial plan from start to finish (the classical "waterfall" style), the problem is broken down into minimal milestones which can each be evaluated when complete. This means that the project is continuously being tested and considered by all stake holders. The result of this is that development is given many opportunities to adapt in the case that problems arise, or if the program does not match user needs, or if those needs change, or any other such issue.

Applying computational techniques and tools in developing scientific software or to support the work of researchers in other disciplines calls for special attention to the unique qualities of a research environment. It is easy to see superficial similarities that the interdisciplinary research environment holds with a production software development environment. For example, in both cases there can be ongoing changes to what is being built, otherwise known as *requirements creep*. One could say that this is a central advantage of working with software: it can be redesigned and rebuilt without repercussions in the physical world, so it makes sense to take advantage of this flexibility. It is not surprising to see this characteristic in any endeavour that adopts software as a tool. Yet there are fundamental differences in who builds research software and what exactly is being constructed. Despite many varieties of development frameworks available today, they all assume a production model at their core: a product is developed for a client. There is a gap between current methods of software development and the needs of the research environment [10].

The communication problem is a recognized challenge for all software development projects [11, 12]. In interdisciplinary research, this is intensified because all team members are specialists. Researchers spend many years building up their domain expertise and cannot be expected to develop deep understanding of another field over a short period of time. By the very nature of research, the topics under investigation are cutting-edge. Finally, in order for the results of a project to be recognized, it is necessary to be able to communicate the workings of any computational elements to reviewers. For these reasons it is vital that there are methods for expressing the essence of an idea that allow for critical inspection, validation and modification.

Traditionally, a software development process is focused on producing a final product for an end-user. In the context of simulations of social systems, scientific research uses programs as experiments, to test theories and generate ideas that lead to new ones. Testing is useful at each stage of theory development, so a single project can generate a number of programs; they should be thought of as a set of related experiments. In science, the core methodology for discovery of new knowledge can be concisely described as the iteration of the hypothesis-experiment-result-conclusion cycle [13].[1] In the case of the social sciences, this includes exploratory techniques as well as classic inductive and deductive approaches. The programs developed for

---

[1] "A SCIENTIST, whether theorist or experimenter, puts forward statements, or systems of statements, and tests them step by step. In the field of empirical sciences, more particularly, he constructs hypotheses, or systems of theories, and tests them against experience by observation and experiment." [13]

scientific research are computational experiments, and in order to accept the reality of iteration, it is necessary to envision the software developed for a research project as a set of related experiments. In order to relate to one another, they must share the same theoretical foundation. In other words, it is the conceptual schema that holds the continuity between programs, and not their functionality. However, the specific implementation of any program will be determined by the requirements of the experiment it embodies. Furthermore, since the software being developed is for the use of the team members themselves, the final configuration of that software is not a primary concern. The configuration of a program is only a concern to the extent that team members find it usable. These characteristics firmly shift the focus of the development cycle to the design or prototyping phase instead of implementation [14].

Central to all of these unique characteristics of the research environment is the underlying aim of scientific discovery. A completed program is not the final goal: software is used as a route to testing existing theories and creating new ones. Software is important because of the results experiments can provide. It can act as a sandbox where scientists are able to try out different ideas. It is also important because the transformation of theory into something computable captures the concepts being considered in a mathematical form. This mathematical model provides a blueprint for researchers to explain their ideas, or for peers to analyze its validity. It allows different projects to be compared to one another, and it acts as a foundation for further exploration of the subject matter. This process can be assisted by the adoption of current best coding practices [15]. Agile methods are well suited to research, particularly the importance placed on people and the idea of iterative development [11]. However, new techniques tailored specifically for the research environment can help to encourage software development that is more successful in generating new knowledge.

It is important for all researchers involved in the project to be able to understand and identify how the computational model works, or in other words, what is going on "under the hood". This does not mean that all members should know programming, however, their understanding of the model will allow them to recognize issues that need to be addressed and either make the alterations directly (through a UI perhaps) or be able to communicate directly to the development members. This is especially important with an experimental project, since vital characteristics may change with each experiment, and may diverge significantly from the design agreed upon in initial phases. Clear representation is necessary for this, and graphical feedback can provide a satisfying way of achieving this. It is important to remember that the goal of any such graphical output is to communicate the behaviour of the program, and not to give an illusion of validity or to otherwise hide inner mechanisms. It also will make it easier to confirm the validity of the computational model. A program with obscure innards is useless to a researcher: they have no confidence in its results (it is just bells and whistles to them), and it cannot be used academically, because the source of any results it gives cannot be explained, so it is not convincing to peer readers.

4 Roles of Interdisciplinary Research Software

The roles computational modelling can play in a simulation modelling research initiative are outlined here. It is upon these points that the framework presented in the next section is premised.

Reinforce Iterative Discovery The executable nature of software means that it can be tested for various things such as accuracy and bugs whenever a change is introduced. Feedback from such testing can be used to iteratively improve either the program or the model itself. Of course, this is also possible with non-computational research, but software encourages this feedback loop through interactivity. To maintain this process, it is important for the mathematical model and computational models to remain flexible and capable of change.

Visualization Visual representation of data allows human users to examine and interpret using natural cognitive and pattern-matching capabilities, in addition to deductive reasoning. Ideally, a variety of visualizations should be available, allowing the user to alter their view of the data in order to consider different perspectives or subsets of the entire data set, which can be overwhelmingly rich in a social simulation.

Formalization Computational modelling forces researchers to precisely define the modelled elements of the subject phenomena so that it is possible to execute it on a computer. This process helps to frame existing domain knowledge in a mathematical manner. It also highlights areas which are poorly understood (which can become clear since they are difficult to formalize) and areas which are not important to the current research focus (which can become clear when defining them takes more effort than the value they add to the model).

Testing Ideas Computational models are ideal for testing hypotheses in a sandbox-like environment.

Raising Questions The modelling process can help raise questions that can be answered by more traditional research in the domain field. This can happen when information is missing to complete the design of the mathematical model, or if the computational model produces contradictory or conflicting results. In these cases, further literature review or discussion among experts may help to elucidate the missing information. It can also guide field research by suggesting questions to ask in future surveys.

Demonstration A computational model is valuable if it can illustrate domain concepts to non-experts. The dynamic qualities of a simulation combined with visualization capabilities combine to provide a powerful means of explaining complex phenomena in a straightforward manner.

Communication of Ideas As a formal model of a concept, the computational model is a proposed theory that can be used to communicate new ideas in an interactive manner among peers. Likewise, the model should be transparent enough that peers can examine it critically to point out problems, ask for clarification, or use the ideas in other research.

Complex Calculation A computational model is often a combination of many simple rules and concepts, but in aggregate the behaviour of such a model can easily be beyond the comprehension or predictive capabilities of a human expert. In this way, a simulation program can act in the place of a human expert, correctly combining rules and concepts to produce aggregate (and sometimes emergent) results.

Reproducibility A hallmark of the scientific process, since it allows results to be validated independently of the original claimant. Computational models are reproducible in two senses: one, in a very literal sense, the simulation program can be distributed and executed by other people, demonstrating the results directly to them (or not, depending on their independent analysis); and two, in that aspects and elements of the proposed model can be taken out and used in other projects, or modelling ideas from other projects can be combined with or substituted into the proposed model [16]. The result of this latter point is that the model concepts can be tested and validated more widely. From a Popperian perspective, if the model concepts are transferrable and useful, they are more general and qualify for greater esteem and respect (at least until shown to be problematic)[13]. Reproducibility of specific runs can even be achieved for stochastic systems by recording the seed number of the random number generator.

5 Social System Computational Modelling Framework

Here, a high level description of the characteristics and goals of a framework for pursuing computational social science is presented. The framework includes both a process through which the model is developed, and a structure for the simulation software that closely aligns with this approach. Note that this does not refer to a specific piece of software, but rather the structure and features a simulation software should possess in order to fully support the model development process.

The agile software development paradigm is in general the most appropriate for working on computational models. The emphasis on flexibility to change is crucial here, since the level of understanding and current focus of interest is subject to ongoing change. Scientific fallibility means that we must be prepared for the discovery of errors in our model, and be able to try out promising alternatives when they appear. It is important to recognize the role of insight as both a driver and goal of working on this kind of project. When it comes to the model of the system, the attitude of "if it isn't broken, don't fix it" is not appropriate; instead it is a constant drive to improve something that is necessarily broken (to some extent). Another reason for an emphasis on agile methods is their focus on the role of human beings in the development process. Team members bring a variety of skills, knowledge, and needs to any project, and failing to take advantage of the good and/or deal with the difficult can be catastrophic in terms of success of the project. With these ideas in mind, some important characteristics of the software package used to implement the simulation system should be:

1. *Easy to change*, so that suggestions and requests for changes, as well as new ideas, can be implemented in a rapid manner. Obscure programming languages

are problematic here, or complicated libraries the code is dependent upon (e.g., for graphics).

2. *Easy to distribute*, so that each team member can easily remain up-to-date with regards to the latest build and test it out on any machine they have access to. Software that is difficult to compile or that requires specific operating systems or settings may be intimidating to some team members, and prevent adoption.

3. *Easy to use*, so that all team members feel comfortable using it to test out their ideas. This can accomplished in part by simplying the user interface and layout of the program. It can also be accomplished by identifying and maintaining an effective user interface, so that users are not forced to relearn the program frequently. Changes to the user interface are acceptable depending on the level of their contribution to the program.

4. *Easy to discuss*, so that the experiences and insight of users can be efficiently converted into improvements to the software. This requires a user-friendly system for reporting errors and discussing ideas. Time spent on establishing a shared vocabulary to describe important features and concepts is also valuable.

The structure of this framework is based on the widely accepted hypotheticpo-deductive model of the scientific method [17]. Generally, this is broken down into a series of stages applied iteratively. For example: observation, hypothesis, methodology, testing, and analysis. In principle, all stages of this process should be represented in a simulation software package, since this allows researchers to tighten the iterative cycle of experimentation. They should not have to waste time switching back and forth between software packages and data formats. Of course, they do not need to visit every step of the cycle each time: they may be uninterested in results analysis at an early stage of sensitivity analysis, where they are interested in parameter values and are still becoming familiar with the qualitative features of model behaviour.

Expanding upon the iterative computational model development process first presented in [18], the stages of the experimental process are:

Conceptual Design A general description of the features and expectations of the model. Relevant domain theory and ideas.

Mathematical Model Equations, formulas, and mathematical statements used to precisely define the structure of the model.

Computational Model (Program Code) The parts of the program related to the simulation model (not technical aspects like the GUI, visualization, etc.) should always be visible to users. Having it visible at all times would encourage the writing of clear code, and allow users familiar with the programming language to verify that the computational description of the simulation entities matches their expectations. It also encourages peers to check over the relevant code while testing out the software, without forcing them to sift through the entirety of the code base. A straightforward specification language (e.g., Abstract State Machines [19]) is ideal for this purpose.

Experimental Design This is where the initial state of a system and the duration of an experiment are specified. Combined with the user's expectations, this composes the scientific claim or hypothesis of this experiment.

Playback Here, the user can view the behaviour of the system as it passes through time during the simulation. They are capable of shuttling back and forwards through time to compare relevant states, such as the beginning, midpoint, and end. Options for viewing all simulated variables and entities should be accessible here.

Visualization/Analysis The data generated by the experiment is presented here for analysis. Currently, the changes in predetermined variables over time are plotted here, but this stage could benefit greatly from including a visualization suite (pre-existing or custom) that would give users the freedom to modify their view of the data as they consider it. In particular, it would be most useful if the visualization tool was capable of identifying all of the variables of the entities included in the simulation, so that data options at this stage would update automatically whenever the model is reprogrammed.

Each of these stages can be clearly included in the design of the software framework for working on social system simulations, for example, as a tab in software user interface. It may be a good idea to limit some users from changing some elements,

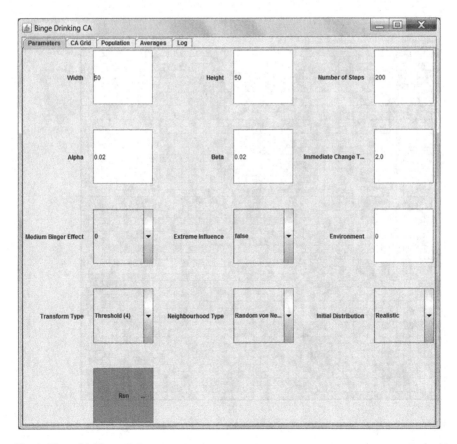

Fig. 1 Binge drinking cellular automata program: parameter set-up

particularly the model code, but it should still be available for them to check over. In the case of something like the conceptual model, a box of text is likely sufficient: it would contain expectations, ideas, and references to relevant literature (perhaps in the form of hyperlinks). Having this as a reference is useful for some or all of the people using the program, even if the project has moved beyond frequent updates of this aspect.

Figures 1, 2, and 3 show the Binge Drinking Cellular Automata program, a model of peer influence on the drinking behaviour of undergraduate students [20]. This model was one of several projects whose development led to the framework presented here [18, 21, 22]. As shown in the screenshot, the program interface has the following tabs (which correspond to the stage in parentheses): Parameters (experimental design), Simulation (playback), Population (visualization), and Averages (visualization). In this case, there is also a Log tab which contains system output (including random number generator seed values). It is important that the software reinforce good scientific practices, and remain a unifying theme across different

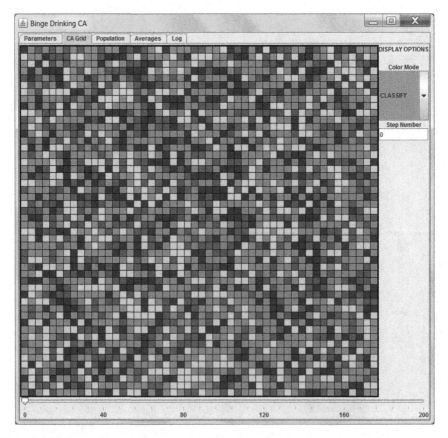

Fig. 2 Binge drinking cellular automata program: simulation playback

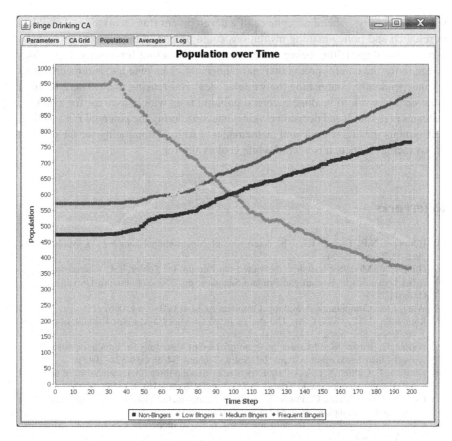

Fig. 3 Binge drinking cellular automata program: population visualization

projects, ideally across different work groups, so that they may more easily understand each other's work. Indeed, developing and maintaining a clear and accessible user interface will allow users to get to understand it better so that they are less intimidated by the software and are able to focus more of their attention on the computational model it is simulating.

6 Final Words

The framework presented here for the construction and development of computational models of social systems is the result of collaborative work on several computational modelling projects in the social sciences. It succesfully brings the varied elements required for producing a computational model together in a way that emphasizes the integrity of the model and facilitates feedback-driven improvement. Perhaps more importantly, general acceptance of a development framework by stakeholders in research and policy making institutions could be one of the factors that leads to computational modelling being more widely used as a tool for inquiry. It provides structure for the careful production and refinement of knowledge, and mechanisms for independently establishing the value of ideas and reusing them. There is still a great deal of work to be done before simulation is as widely used as, for example, statistics in research and decision making, but considering the potential for a method that exploits the algorithmic and mathematical nature of computing for the purpose of advancing science, it is a worthwhile goal to aim for.

References

1. Hummon, N.P., Fararo, T.J.: The emergence of computational sociology. J. Math. Sociol. **20**(2–3), 79–87 (1995)
2. Gilbert, N.: Modeling sociality: the view from Europe. In: Kohler, T.A., Gummerman, G.J. (eds.) Dynamics in Human and Primate Societies, pp. 355–371. Oxford University Press, Oxford (2000)
3. Wing, J.M.: Computational thinking. Commun. ACM **4**(3), 33–35 (2006)
4. Ramanath, A.M., Gilbert, N.: The design of participatory agent-based social simulations. J. Artif. Soc. Soc. Simul. **7**(4), (2004)
5. Fortino, G., Russo, W.: Eldameth: an agent-oriented methodology for simulation-based prototyping of distributed agent systems. Inf. Softw. Technol. **54**(6), 608–624 (2012)
6. North, M.J., Collier, N.T., Vos, J.R.: Experiences creating three implementations of the repast agent modeling toolkit. ACM Trans. Model. Comput. Simul. **16**(1), 1–25 (2006)
7. Luke, S., Cioffi-Revilla, C., Panait, L., Sullivan, K., Balan, G.: Mason: a multiagent simulation environment. Simulation **81**(7), 517–527 (2005)
8. Gilbert, N., Terna, P.: How to build and use agent-based models in social science. Mind Soc. **1**, 57–72 (2000)
9. Leombruni, R., Richiardi, M., Revelli, L., Studies, C.E., Real, V.: A common protocol for agent-based social simulation. J. Artif. Soc. Soc. Simul. **9**(1), (2006)
10. Kelly, D.F.: A software chasm: software engineering and scientific computing. IEEE Softw. **24**, 119–120 (2007)
11. Fowler, M.: The new methodology. http://martinfowler.com/articles/newMethodology.html (2003)
12. Berry, D.M.: Formal methods: the very idea–some thoughts about why they work when they work. Sci. Comput. Program. **42**(1), 11–27 (2002)
13. Popper, K.: The Logic of Scientific Discovery. Hutchinson Education, London (1959)
14. Rickett, C.D., Choi, S., Rasmussen, C.E., Sottile, M.J.: Rapid prototyping frameworks for developing scientific applications: a case study. J. Supercomputing **36**, 123–134 (May 2006)
15. Baxter, S.M., Day, S.W., Fetrow, J.S., Reisinger, S.J.: Scientific software development is not an oxymoron. PLoS Comput. Biol. **2**, e87 (2006)
16. Buckheit, J., Buckheit, J.B., Donoho, D.L., Donoho, D.L.: Wavelab and Reproducible Research, pp. 55–81. Springer-Verlag (1995)
17. Godfrey-Smith, P.: Theory and Reality: An Introduction to the Philosophy of Science. University of Chicago Press, Chicago (2003)
18. Brantingham, P., Glsser, U., Jackson, P., Vajihollahi, M.: Modeling criminal activity in urban landscapes. In: Memon, N., David Farley, J., Hicks, D.L., Rosenorn, T. (eds.) Mathematical Methods in Counterterrorism, pp. 9–31. Springer, Vienna (2009)
19. Börger, E., Stärk, R.F.: Abstract State Machines. A Method for High-Level System Design and Analysis. Springer, Berlin (2003)

20. Jackson, P., Reid, A., Huitson, N., Wuschke, K., Dabbaghian, V.: Drinking with friends: a cellular automata approach to modeling peer influence of on binge drinking behaviour. In: International Symposium on Cellular Automata Modeling for Urban and Spatial Systems (CAMUSS), pp. 147–157 (2012)
21. Dabbaghian, V., Jackson, P., Spicer, V., Wuschke, K.: A cellular automata model on residential migration in response to neighborhood social dynamics. Math. Comput. Model. **52**(9–10), 1752–1762 (2010)
22. Pratt, S., Giabbanelli, P., Jackson, P., Mago, V.: Rebel with many causes: a computational model of insurgency. In: IEEE International Conference on Intelligence and Security Informatics, pp. 90–95 (2012)

Chapter 3
Modelling Epistemic Systems

André C. R. Martins

Abstract In this Chapter, I will explore the use of modeling in order to understand how Science works. I will discuss the modeling of scientific communities, providing a general, non-comprehensive overview of existing models, with a focus on the use of the tools of Agent-Based Modeling and Opinion Dynamics. A special attention will be paid to models inspired by a Bayesian formalism of Opinion Dynamics. The objective of this exploration is to better understand the effect that different conditions might have on the reliability of the opinions of a scientific community. We will see that, by using artificial worlds as exploring grounds, we can prevent some epistemological problems with the definition of truth and obtain insights on the conditions that might cause the quest for more reliable knowledge to fail. A simple model of scientific agents opinions influenced by colleagues and experimental results about the world they live in will also be discussed.

1 Introduction

The classical description of the work of a scientist is one that many of us would like to believe true. According to it, scientists are as close as reasonably possible to believe to selfless individuals who pursue knowledge using always the best means available to them. They propose hypothesis and theories according to what they feel would describe all relevant data better. And, from those models, they draw predictions that are tested under very strict conditions. Finally, still according to our idealized view, it is the agreement between observations and those predictions that dictate which theories are better accepted. Conditions of beauty, like simplicity and symmetry, can be invoked when deciding between equally well suited theories, depending on who is telling the tale. But that is it, and no other considerations should be included.

André C. R. Martins (✉)
GRIFE-EACH, Universidade de São Paulo, R. Arlindo Bettio, São Paulo–SP, 1000, Brazil
e-mail: amartins@usp.br

V. Dabbaghian and V. K. Mago (eds.), *Theories and Simulations of Complex Social Systems*, Intelligent Systems Reference Library 52,
DOI: 10.1007/978-3-642-39149-1_3, © Springer-Verlag Berlin Heidelberg 2014

Obviously, real life is not so simple and scientists do defend their own ideas because it is their own or because they like it better. And, as in any human activity, one will find individuals whose agenda is based only in self-interest. However, as long as community moves in the right direction and eventually corrects any mistakes, we could, in principle, consider these problems as just perturbations to an otherwise reasonably precise description of the scientific enterprise. In particular, the amazing advances of knowledge in many areas, where old ideas have been replaced by newer and better ones, seems to lend a lot of credit to that idealized description. At the very least, as a good approximation for the system as a whole, even if it fails in individual cases.

And, indeed, we do have controls to avoid the more serious problems. Replication (or reproducibility) [1–3] is considered fundamental by most and with good reason. Once an independent researcher obtains the same results, chances of error in the first report are obviously smaller. But, most important of all, if the groups are truly independent and the second group has nothing to gain from either a positive or a negative result, the chances that the first group published fraudulent results basically disappear. This makes the problem of self-serving scientists much less important than it might otherwise be. It also points out to another central feature of the scientific endeavor that is sometimes neglected. That is, the social aspect of Science.

As in many human activities, the outcomes of scientific research are nowadays a social product of a community of scientists. Theory and measurement are often done by different individuals, except in cases where just a fact, a simple idea, or a property of some material or drug are tested. Even in these cases, there is often a group behind the results, as many fields have become so complex that the expertise of many different researchers is often needed. Main examples of this are the "Big Science" projects, such as the mapping of the human genome or the hunt for the Higgs boson. You have different people taking care of different aspects of the problem, some building the equipments, others operating them, a group specializing in collecting the data, more people to store it properly, others to interpret it and so on. Even in smaller projects, like a simple simulation that one can program alone, we often use software developed by a third party, software we trust to do what we are told it does. We depend on the quality of the information and work performed by others in most of the research done nowadays. And, while it might be reasonable, under some circumstances, to assume that most of the information we use were produced in a competent and honest way, it is certain that we can not claim it for all the information [4, 5]. Despite what we are taught in school, arguments of authority are actually unavoidable [6] and we have to understand the effects this might have.

As soon as we realize that we have come to depend on the results of others even to make our own research, we have to realize we might risk running into epistemic trouble. While it is clear that some social aspects are clearly positive (as, obviously, replication), social influence could lead some people away from the best answer. That means that understanding which social aspects can help the reliability of a result and which ones are more likely to be detrimental is an important task. And we must acknowledge that there are different accounts of the scientific process that do not describe it in such beautiful collors, like, per example, the Strong Program in the Sociology of Scientific Knowledge [7–10]. That program defends that Science is just another human social activity. And, as such, not more reliable than any other descriptions of the world obtained by other methods. While scientific advances seem to contradict that in a very strong way, one must admit that some of the criticism might be correct and that there might be ways to make the whole scientific activity even more reliable and less prone to error. In a time where no single human can check the correctness of his whole field alone, understanding the consequences of social interaction and errors can be fundamental to the future quality of the Science we will make.

Furthermore, there are fields in which published results have often been consistently shown to be wrong [11–13], exactly due to the way the Science and publications are structured. Biases towards publishing only some types of results exist and are very prejudicial. It is also already known that replication is a far less common practice than it would be expected in areas as different as Marketing [14, 15] and Medicine [16]. These findings make it very clear that it is imperative that we study and understand well the effects of how the way Science is structured and how scientists interact might affect the quality of our conclusions.

Of course, trying to figure how reliable an answer is or if we can really say we know something about the world is an old philosophical problem that we will not answer so simply. However, the question of reliability of results might be particularly well suited to be analyzed with the tools of simulation and agent based models. Simulating a community of scientists will not answer the deepest problems of Skepticism [17], like questions about the existence of a real world or our ability to trust our sensory information about it. It wouldn't be impossible to devise simulations where the agents face those problems and try to learn whether they observe is real or not, but that is not the intent here. Even assuming that it is possible to know something about the world, we still can learn a lot about different ways to learn. Agent based models of Science can help us explore which strategies and behaviors are more likely to cause problems or to get us closer to the best answers. This happens not only because we can explore unforeseen consequences of interaction rules that we wouldn't see in models built using just human language, but also because we can, as programmers, choose which theory is the best one in the artificial world of our agents. And therefore, in principle, check the effects of different social structures on the choice of the best theory. Of course, any limitations that apply to current models of social interaction will apply here fully and the results must be analyzed with the typical doubt we associate with tentative new ideas. But we can get some advice on where it might be more likely that a change in current practices will lead to better outcomes.

1.1 Existing Models for Describing Scientific Practices

In this chapter, I defend one specific use for agent models, one that has seen a small number of papers up to now. However, it is interesting to acknowledge that the applications of computational and mathematical modeling to communities of scientists is an area that has been seen a steady increase in the interest and number of papers. Different researchers have dedicated themselves to different problems, most of them tending towards describing the society of scientists. Therefore, it makes sense to provide a very brief and very incomplete review of those applications. Those works can be divided in three major areas: methods to measure the quality and impact of a scientist work; studies of the structure of Science, as per example, the networks of authors and their papers; and the modeling of knowledge, either just looking at its diffusion, or at the acceptance of new and better knowledge. It is particularly worth noting that recently, there was a full edition of the Journal of Artificial Societies And Social Simulation [18] as well as a book [19], both dedicated to simulation of Science.

Measuring the impact of the work of each of us can be an important task. It has significant influence on our careers and scientific policy makers could use such this information when deciding how to divide the funding [20]. It is no surprise therefore that attempts at providing quantitative measurements for the relevance of a scientist work have become popular and several different proposals now exist. They include the now unavoidable (despite the biases it introduces) H-index [21], initially proposed to avoid the problems of measurement brought by total number of citations, that could be a very large number even if the scientist had just collaborated in one very well cited article. Several other measurements were also proposed [22, 23] and many others are probably being investigated. One interesting aspect of the problem that still needs to be addressed, however, is the impact these indexes have in scientific production. Scientists are usually reasonably intelligent individuals and, therefore, they might change the way they work in order to have better values at the measurement they are evaluated with, instead of worrying only about making better Science.

On the evolution of scientific fields, the ability to measure and quantify production can be both important and interesting in itself [24]. The existence of big databases of published material has allowed a large number of studies on structure that is formed by articles and researchers. There are models describing the networks that are formed when scientists cite others [25–29] as well as the networks of co-authorship [30–32], and even of articles connected by the same specific area [33]. This mapping effort allows us to have a much clearer view of how Science is done. Also, the existence of these studies and the data they are based on provides us with indirect measures of social influences among researchers. A complete agent model of scientific activities should either use or explain the structures that were observed; however, we are still far from obtaining such a model, the same way that we are far from having a really reliable and complete model of how people influence each other choices in any field of human behavior. Existing models, as we will see, are still simplistic versions built to address just a few questions and not really general enough.

Still, more recent years have seen the appearance of the first models of idea diffusion in scientific communities, including approaches that make use of the Master equation [34] and population dynamics [35] instead of agent models. Among the agent models, several just try to describe the statistical features observed in the studies that measured number of papers and how they are connected by proposing simple rules of influence [36–39] such as, per example, copying the citations of read papers. Attempts at introducing more sophisticated agents, with better cognition, also exist [40]. Many general ideas not completely developed have also been proposed as possible basis for future models [41–43], including general ideas about the practice inspired by the works of philosophers of Science [44].

Of particular interest to the view defended here are models for the diffusion of ideas and opinions that try to answer whether our current practices are good enough or whether we should work to change some of them. One first model used the Bounded Confidence model [45, 46] of Opinion Dynamics [46–52] as basis, exploring if agents would approach a true value of a parameter given that some of the agents were truth-seekers, who had a tendency to move towards the correct result [53]. However, by adopting a purely continuous version of opinion, that model, while very interesting, is not enough to describe the choice of theories. The effects of peer review [54, 55] were also suggested as an important field to explore by the use of models, since many of the proposals for change in the peer review process have not been explored with the necessary depth. In the same line, discussions of the possible use of modeling to make Science more reliable already exist [56]. However, a strong push forward in developing these models is still to be observed.

Another promising line of investigation on the reliability of Science is the study of the effects of the existence of agents who make claims that are stronger than the actual observations warrant. The main question explored there is if this type of lie could help in the convincing of general public [57, 58]. Finally, I have also proposed a model to explore acceptance of better theories [59], one that was able to show that the idea that Science advances thanks to the retirement of old scientists [60] might be true. In the Sect. 3, we will discuss it and I will show why I believe it can provide the basis for a more sophisticated and maybe even a little less artificial description of the scientific enterprise. That study was based on an Opinion Dynamics model that distinguishes choices and strength of opinion [51, 61] and is also based on notions of rationality. A model working on the same problem, where actual publications were introduced as the means that scientists use to communicate their results was also recently proposed with some very promising results [62].

2 Epistemology and Modelling

Before presenting a model, it makes sense to discuss a few issues regarding the applicability of mathematical and computational modeling to the epistemic problem of finding better ways to obtain reliable knowledge. The first question worth mentioning, but too complex to be discussed properly here, is exactly the meaning of knowing [63]. Traditional accounts define that one knows something when the person has a justified true believe about that something, but there are known problems with that definition [64].

As mentioned above, one interesting feature of artificial worlds is that, since the programmer is playing the part of a creator god, we can circumvent, for the agents, any of the arguments of Skepticism about the possibility of knowing something about the world. Even softer skeptical arguments, like the problem of induction [65, 66] can be either dismissed by constructing a world where the rules don't change, or, eventually, one could investigate what would happen if the future did not exactly follow the same laws as the past. By controlling the laws of the artificial universe, different scenarios and their impact on the learning of the agents can be much better investigated than in the real world, where we can only know our theories, but not be always certain about which of them are really better. Of course, this means developing better models for the dynamics of scientific opinions, but the possibility that those models can be quite helpful at exploring epistemic problems should be clear.

In the next Section, I will describe a model that I believe has the potential to be built upon so that we can arrive at a more realistic description of the problem. But, before that, it is good to debate the fundamental ideas behind it and how they relate to current developments in Epistemology. One line of reasoning in Epistemology claims that all we can really know about the world are probabilities that our ideas are correct. When making any new observation, those probabilities must change according to what we have learn, by obeying the Bayes Theorem. This idea can be called Bayesian Epistemology [67–70]. This is a normative theory, in the sense that it claims that this is the best way to make inferences, and not necessarily the way we actually perform them. The claim that scientist behave and choose the theories they support in a way similar to the specifications of Bayesian methods is known in Philosophy as Confirmation Theory [69, 71–75]. While it is known that humans are actually bad probabilists [76], there is some evidence that scientists do not deviate so much from a Bayesian point of view [77]. Interestingly, not only scientists, but there is a growing body of evidence that the way people reason about problems and theories is actually close to Bayesian analysis [78, 79], if not exactly so. The so-called error in human probabilistic reasoning can actually be explained by assuming that our brains work with more complex models of reality than what was supposed in the laboratory tests [80, 81].

Therefore, we can use Bayesian models as basis for the behavior of our agents [82], even if only as a reasonable first approximation. Such an approach has already produced a number of Opinion Dynamics models [51, 52, 59, 61, 83–91] and it was also used to show why a non-specialized reader should only come to trust a scientific model in cases where there might be errors and deception from the part of the authors, when the results are replicated by a third party [2, 3, 92].

3 A Model for the Spread of a New Theory

3.1 The Model

Assume each agent must choose between two theories, A and B. This choice is a discrete choice between those alternatives, but agents can have a strength of opinion, represented by the probability p_i that each artificial scientist i assigns to the possibility that A is the best description. If $p_i > 0.5$, agent i believes A is the best option, otherwise, it is B. This model is basically the "Continuous Opinions and Discrete Actions" model (CODA model), previously used to explore the emergence of extremism in artificial societies [51, 61]. CODA is based on the idea that agents believe that, if A is the best choice, each of the neighbors of an agent, located in a given social network [93, 94], will have a probability $\alpha > 0.5$ of choosing A (and similarly, for B). In this context, it is easier to work with a transformation of variables, given by the quantity

$$v_i = \ln \frac{p_i}{1 - p_i}.$$

Here, if $v_i > 0$, we have $p_i > 0.5$ and, therefore, a subjective belief in favor of A; if $v_i < 0$, the agent chooses B. By applying Bayes Theorem, we obtain a very simple update rule for v_i, when agent i observes the choice of its neighbor j, given by

$$v_i(t+1) = v_i(t) + \text{sign}(v_j) * a,$$

where a is a step size that depends on how likely the agents believe it is that their neighbors will be correct, that is, it is a function of α. If we renormalize the update rule, by using $v_i^* = v_i/a$ instead, we will have

$$v_i^*(t+1) = v_i^*(t) + \text{sign}(v_j^*), \tag{1}$$

making it clear that the value of a is irrelevant to the dynamics of choices, since that dynamics depends only on the signs of v_i (or v_i^*). The update of the opinions is asynchronous, with one agent and its observed neighbor randomly drawn at each iteration. Time is counted as multiples of the number N of agents, so that, when time increases by one, each agents has interacted once, in average.

In order to introduce the influence of observations in this problem, a proportion τ of the scientists actually perform experiments. Whether an agent is an experimenter or not is decided at the beginning of each realization for each agent and they do not change their status. This experiments can have a stronger or weaker influence on this agents, measured by the probability ρ. Here, ρ stands for the chances that, at each time step, if the drawn agent is an experimenter, it will observe the Nature, instead of being influenced by a neighbor. At this point, we assume experiments always provide the same answer, agreeing with the correct choice A. We also make it so that the agent will update its opinion by the same amount it would from an interaction with a neighbor. That is, opinion at each step is always updated by 1, according to Eq. (1), either following the Equation exactly when social influence happens, or by adding 1 in favor of A, when Nature is observed. Notice that a stronger experimental effect can easily be introduced simply by allowing ρ to be larger and that means no new parameter needs to be introduced here.

3.2 Results

It was observed that, unless τ is reasonably large, even when experimentalists are very weakly influenced by social effects (ρ close to 1.0), they fail to convince the whole population and clusters of agents preferring the worst theory survive. This effect was particularly strong when theory A was new and, therefore, started with a small proportion of supporters, indicating that the old view would survive simply from social effects and the strength of opinion of its old supporters. Interestingly, when retirement was introduced in the system, with agents replaced by new ones with moderate initial opinions, it became much easier for new and better ideas to invade, confirming Kuhn's notion that Science would advance and accept new paradigms due to the death of the old scientists [60].

Of course, those results were obtained with a very simplified version of the scientific activity. This was intentional, in the old Physics tradition of starting at the very simplest model one can imagine and later add new features, as they become needed. The idea is to understand which details are responsible for which features of the whole problem. As such, expanding the model from its several approximations is a natural next step. Per example, in the model, social influence happened only between peers and they were all located on a non-realistic square lattice. No publications existed in the model and Nature always gave the same answer, meaning one theory was clearly superior to the other. And, while people would reinforce their ideas given a certain neighborhood, all agents were supposed to be honest, in the sense that they would always choose the theory they really believed more likely to be truth. Despite all this, the model is very easy to expand. By changing the likelihoods agents assign to the choices of others, we could introduce more realistic versions of social influence. Researchers could do papers, adding a new layer to the problem, and the quality of those papers could depend on the correctness of their argument. And, of course, more realistic networks can be trivially introduced.

In order to illustrate these differences, new cases for small-world networks were run specifically for this chapter. The results are shown in Fig. 1. All cases shown

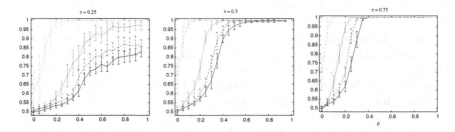

Fig. 1 Proportion of agents aligned with the external field (correct scientists) as a function of ρ. Every realization had initial conditions where each agent had 50% chance of supporting either theory. The *solid line* corresponds to $\lambda = 0.0$, the *dashed* one to $\lambda = 0.1$, the *dotted* one to $\lambda = 0.25$, and the *dash-dotted* one to $\lambda = 0.5$. (*Left panel*) $\tau = 25\%$. (*Middle panel*) $\tau = 50\%$. (*Right panel*) $\tau = 75\%$

correspond to a initial square lattice with 32^2 agents, where each link was initially randomly rewired with probability λ and kept fixed after. Each point is the result of the average of 10 different realizations of the same problem and the bars correspond to the standard deviation of the observed values. Initial conditions were such that at first both theories had equal amount of support and the system was left to interact for $t = 200$ average interactions per agent. Each panel shows the results for different values of λ for a specific proportion τ of experimenters. We can see that as the small-world effect becomes more important (larger rewiring λ), it becomes easier for the system to reach agreement on the best theory. However, the problem with the continuing existence of groups supporting the worse theory remains for the cases where the importance of experiment is not large enough, confirming, at least, qualitatively, the results previously obtained.

4 Conclusion

We have seen that mathematical and computational modeling of scientific agents can be used as a tool to help us understand under which conditions scientific results are more reliable. While it is quite true that the models so far proposed are very simple, we have seen that they can already capture some of the features of the scientific enterprise, such as the need for solid and reliable experimental data. If experiments are not very convincing, we would be in the situation where ρ is small and, therefore, social influence can indeed become the major force in the system, as suggested in the Strong Program. However, as soon as experiments become more important and we have a better connected world, the case for the better theory becomes strong enough to convince everyone. We should also notice that the same model was able to show the importance of the retirement of scientists, as a way to allow new better ideas to spread more easily.

As such, I would like to encourage the modeling community to investigate models for scientific activity. And by that I mean not only from the sociological (and also interesting) point of view of measuring our activities, but also from the point of view of suggesting better practices. While doubt will remain when a suggestion comes from a model, the tools we have make the exploration of the consequences more reliable than simple human intuition and spoken arguments. Models of Science can and should help us make research an even better tool for understanding the world.

Acknowledgments The author would like to thank Fundação de Amparo à Pesquisa do Estado de SãoPaulo (FAPESP), under grant 2011/19496-0, for the support to this work.

References

1. Giles, J.: Nature **442**, 344 (2006)
2. Palmer, V.: J. Artif. Soc. Soc. Simul. **9**(1), 14 (2006)
3. Martins, A.C.R.: JASSS-J. Artif. Soc. Soc. Simul. **11**(4), 8 (2008). http://jasss.soc.surrey.ac.uk/11/4/8.html

4. Check, E., Cyranoski, D.: Nature **438**, 1056 (2005)
5. Cho, M.K., McGee, G., Magnus, D.: Science **311**, 614 (2006)
6. Hardwig, J.: J. Philos. **82**(7), 335 (1985)
7. Barnes, B., Bloor, D.: Relativism, Rationalism, and the Sociology of Knowledge. In: Hollis, M., Lukes, L. (eds.) Rationality and Relativism, pp. 21–47. MIT Press, Cambridge (1982)
8. Pickering, A.: Constructing Quarks: A Sociological History of Particle Physics. Edinburgh University Press, Edinburgh (1984)
9. Latour, B., Woolgar, S.: Laboratory Life: The Construction of Scientific Facts. Princeton University Press, Princeton (1986)
10. Bloor, D.: Knowledge and Social Imagery. University of Chicago Press, Chicago (1991)
11. Ioannidis, J.P.A.: PLoS Med **2**(8), e124 (2005)
12. Begley, C.G., Ellis, L.M.: Nature **483**, 531–533 (2012)
13. Sarewitz, D.: Nature **485**, 149 (2012)
14. Hubbard, R., Armstrong, J.S.: Int. J. Res. Mark. **11**, 233 (1994)
15. Evanschitzky, H., Baumgarth, C., Hubbard, R., Armstrong, J.S.: J. Bus. Res. **60**, 411 (2007)
16. Ioannidis, J.P.A.: J. Am. Med. Assoc. **294**, 218 (2005)
17. Klein, P.: Skepticism. In: Moser, P.K. (ed.) The Oxford Handbook of Epistemology, Oxford University Press, Oxford (2002)
18. Edmonds, B., Gilbert, N., Ahrweiler, P., Scharnhorst, A.: J. Artif. Soc. Soc. Simul. **14**(4), 14 (2011)
19. Scharnhorst, A., Börner, K., van den Besselaar P. (eds.): Models of Science Dynamics: Encounters Between Complexity Theory and Information Sciences. Understanding Complex Systems. Springer, Berlin (2012)
20. Yilmaz, L.: J. Artif. Soc. Soc. Simul. **14**(4), 2 (2011)
21. Hirsch, J.: PNAS **102**, 16569 (2005)
22. Egghe, L.: Scientometrics **69**, 131 (2006)
23. Galam, S.: Scientometrics **89**(1), 365 (2011)
24. Egghe, L., Rousseau, R.: Introduction to Infometrics: Quantitative Methods in Library, Documentation and Information Science. Elsevier, Amsterdam (1990)
25. Garfield, E., Sher, I.H., Torpie, R.J.: The Use of Citation Data in Writing the History of Science. Institute for Scientific Information, Philadelphia (1964)
26. Börner, K., Maru, J.T., Goldstone, R.L.: Proc. Natl. Acad. Sci. U.S.A. **101**, 5266 (2004)
27. Radicchi, F., Fortunato, S., Markines, B., Vespignani, A.: Phys. Rev. E **80**, 056103 (2009)
28. Quattrociocchi, W., Amblard, F.: Emergence through selection: the evolution of a scientific challenge (2011). [ArXiv preprint: 1102.0257]
29. Ren, F.X., Cheng, X.Q., Shen, H.W.: Physica A **391**, 3533 (2012). [ArXiv preprint 1104.4209v2]
30. Newman, M.E.J.: Proc. Natl. Acad. Sci. U.S.A. **98**, 404 (2001)
31. Newman, M.E.J.: Phys. Rev. E **64**, 016131 (2001)
32. Barabási, A., Jeong, H., Néda, Z., Ravasz, E., Schubert, A., Vicsek, T.: Physica A **311**, 590 (2002)
33. Herrera, M., Roberts, D.C., Gulbahce, N.: PLoS ONE **5**(5), e10355 (2010)
34. Kondratiuk, P., Siudem, G., Holyst, J.A.: Analytical approach to model of scientific revolution (2011). [Arxiv, preprint 1106.0438]
35. Vitanov, N., Ausloos, M.: Models of Science Dynamics. In: Scharnhorst, A., Börner, K., van den Besselaar, P. (eds.) Models of Science Dynamics, Springer, Berlin (2012)
36. Gilbert, N.: Sociol. Res. Online **2**(2), (1997)
37. Simkin, M., Roychowdhury, V.: Annals Improb. Res. **11**(1), 24 (2005)
38. Simkin, M., Roychowdhury, V.: J. Am. Soc. Inform. Sci. Technol. **58**(11), 1661 (2007)
39. Newman, M.E.J.: Europhys. Lett. **86**, 68001 (2009). [Arxiv:0809.0522v1]
40. Naveh, I., Sun, R.: Comput. Math. Organ. Theory **12**, 313 (2006)
41. Doran, J.: J. Artif. Soc. Soc. Simul. **14**(4), 5 (2011)
42. Parinov, S., Neylon, C.: J. Artif. Soc. Soc. Simul. **14**(4), 10 (2011)
43. Payette, N.: J. Artif. Soc. Soc. Simul. **14**(4), 9 (2011)

44. Edmonds, B.: J. Artif. Soc. Soc. Simul. **14**(4), 7 (2011)
45. Deffuant, G., Amblard, F., Weisbuch, T., Faure, G.: J. Artif. Soc. Soc. Simul. **5**(4), 1 (2002)
46. Hegselmann, R., Krause, U.: J. Artif. Soc. Soc. Simul. **5**(3), 3 (2002)
47. Castellano, C., Fortunato, S., Loreto, V.: Rev. Mod. Phys. **81**, 591 (2009)
48. Galam, S., Gefen, Y., Shapir, Y.: J. Math. Sociol. **9**, 1 (1982)
49. Sznajd-Weron, K., Sznajd, J.: Int. J. Mod. Phys. C **11**, 1157 (2000)
50. Galam, S.: Europhys. Lett. **70**(6), 705 (2005)
51. Martins, A.C.R.: Int. J. Mod. Phys. C **19**(4), 617 (2008)
52. Martins, A.C.R.: Discrete opinion models as a limit case of the CODA model. In: Social Physics Catena (No.3), pp. 146–157. Science Press, Beijing (2012). [ArXiv:1201.4565v1]
53. Hegselmann, R., Krause, U.:J. Artif. Soc. Soc. Simul. **9**(3), 10 (2006). Http://jasss.soc.surrey.ac.uk/9/3/10.html.
54. Sobkowicz, P.: Peer-review in the internet age (2010). [ArXiv:0810.0486v1]
55. Squazzoni, F., Takács, K.: JASSS-J. Artif. Soc. Soc. Simul. **14**(4), 3 (2011)
56. Zollman, K.: JASSS-J. Artif. Soc. Soc. Simul. **14**(15), 4 (2011)
57. Galam, S.: Opinion Dynamics, Minority Spreading and Heterogeneous Beliefs. In: Chakrabarti, B.K., Chakraborti, A., Chatterjee, A. (eds.) Econophysics and Sociophysics: Trends and Perspectives, pp. 363–387. Wiley, Berlin (2006)
58. Galam, S.: Physica A **17**(1), 3619 (2010)
59. Martins, A.C.R.: Adv. Compl. Sys. **13**, 519 (2010)
60. Kuhn, T.: The Structure of Scientific Revolutions. University of Chicago Press, Chicago (1962)
61. Martins, A.C.R.: Phys. Rev. E **78**, 036104 (2008)
62. Sobkowicz, P.: Scientometrics **87**(2), 233 (2011)
63. Shope, R.K.: Conditions and Analyses of Knowing. In: Moser, P.K. (ed.) The Oxford Handbook of Epistemology, Oxford University Press, Oxford (2002)
64. Gettier, E.: Analysis **23**, 121 (1963)
65. Hume, D.: An Inquiry Concerning Human Understanding. Claredon Press, Oxford (1777)
66. Howson, C.: Hume's Problem: Induction and the Justification of Belief. Oxford University Press, Oxford (2003)
67. Kaplan, M.: Decision Theory as Philosophy. Cambridge University Press, Cambridge (1996)
68. Kaplan, M.: Decision Theory and Epistemology. In: Moser P.K. (ed.) The Oxford Handbook of Epistemology, Oxford University Press, Oxford (2002)
69. Bovens, L., Hartmann, S.: Bayesian Epistemology. Oxford University Press, Oxford (2003)
70. Talbott, W.: Bayesian Epistemology. In: Zalta, E.N. (ed.) The Stanford Encyclopedia of Philosophy, The Metaphysics Research Lab, (2008). http://plato.stanford.edu/entries/epistemology-bayesian/
71. Earman, J.: Bayes or Bust? A Critical Examination of Bayesian Confirmation Theory. MIT Press, Cambridge (1992)
72. Jeffrey, R.C.: The Logic of Decision. University of Chicago Press, Chicago (1983)
73. Jeffrey, R.C.: Subjective Probability: The Real Thing. Cambridge University Press, Cambridge (2004)
74. Howson, C., Urbach, P.: Scientific Reasoning: The Bayesian Approach. Open Court, Chicago (1993)
75. Maher, P.: Betting on Theories. Cambridge University Press, Cambridge (1993)
76. Plous, S.: The Psychology of Judgment and Decision Making. McGraw-Hill, New York (1993)
77. Press, S.J., Tanur, J.M.: The Subjectivity of Scientists and the Bayesian Approach. Wiley, New York (2001)
78. Tenenbaum, J.B., Kemp, C., Shafto, P.: Theory-based Bayesian Models of Inductive reasoning. In: Feeney, A., Heit, E. (eds.) Inductive Reasoning, Cambridge University Press, Cambridge (2007)
79. Kemp, C., Tenenbaum, J.B., Niyogi, S., Griffiths, T.L.: Cognition **114**(2), 165 (2010)
80. Martins, A.C.R.: Adaptive probability theory: human biases as an adaptation (2005). Cogprint preprint at http://cogprints.org/4377/
81. Martins, A.C.R.: Judgment Decis. Making **1**(2), 108 (2006)

82. Martins, A.C.R.: A bayesian framework for opinion updates. In: AIP Conference Proceedings **1490**, 212 (2012). [ArXiv:0811.0113v1]
83. Banerjee, A.V.: Quaterly J. Econ. **107**(3), 797 (1992)
84. Orléan, A.: J. Econ. Behav. Organ. **28**, 257 (1995)
85. Lane, D.: Is what is good for each best for all? Learning from Others in the Information Contagion Model. In: Arthur, W., Durlauf, S., Lane, D.(eds.) The Economy as an Evolving Complex System I, pp. 105–127. Perseus Books, Santa Fe (1997)
86. Martins, A.C.R.: J. Stat. Mech Theory Exp. **2009**(02): P02017 (2009). [ArXiv:0807.4972v1]
87. Vicente, R., Martins, A.C.R., Caticha, N.: J. Stat. Mech Theory Exp. **2009**, P03015 (2009). [ArXiv:0811.2099]
88. Martins, A.C.R., Pereira, C.d.B., Vicente, R.: Physica A **388**, 3225 (2009)
89. Si, X.M., Liua, Y., Xionga, F., Zhanga, Y.C., Ding, F., Cheng, H.: Physica A **389**(18), 3711 (2010)
90. Martins, A.C.R.: Adv. Appl. Stat. Sci. **2**, 333 (2010)
91. Martins, A.C.R., Kuba, C.D.: Adv. Compl. Sys. **13**, 621 (2010). [ArXiv:0901.2737]
92. Martins, A.C.R.: JASSS-J. Artif. Soc. Soc. Simul. **8**(2): 3 (2005). http://jasss.soc.surrey.ac.uk/8/2/3.html
93. Newman, M., Barabasi, A.L., Watts, D.J.: The Structure and Dynamics of Networks. Princeton University Press, Princeton (2006)
94. Vega-Redondo, F.: Complex Social Networks. Cambridge University Press, Cambridge (2007)

Chapter 4
Modeling Humain Behavior in Space and Time Using Mobile Phone Data

Ana-Maria Olteanu Raimond and Thomas Couronné

Abstract In this chapter we present an overview of the main sources of data coming from mobile phone tracking and models allowing the use of these data. Several issues due to the quality of mobile phone data are explained. In particular, we provide a taxonomy of mobile phone data imprecision and suggest new metrics to estimate the basic properties of displacements are defined: mobility intensity (speed-like measure) and uncertainty.

1 Introduction

Nowadays many works have been carried out about human behaviors modeling and analysis based on digital tracks generated by the user activity himself. This brand new approach of behavioral sciences via big data is relevant to address and unfold macro processes, and help a deep understanding of the revealed processes.

Human mobility inference from mobile phone data has become popular in the past few years [1–5]. In a short period, many advances have been achieved, based on always bigger datasets.

In this work, we expose the potential of the human activity modeling based on mobile phone records, together with limits and issues raised by this approach. This chapter first describes examples of applications and then provides data sources and acquisition. Finally it exposes the limits and the perspectives of these approaches focusing on the uncertainty and imprecision inherent to mobile phone data.

A. O. Raimond (✉) · T. Couronné
SENSe Orange Labs, Networks and Carriers, 38-40 rue du Général Leclerc, Issy-les-Moulineaux cedex9 92794, France
e-mail: ana-maria.raimond@ign.fr

T. Couronné
e-mail: thomas.couronne@orange.com

2 Models

In this section, we are going to focus on human mobility modeling, considering that human communication, modeled as a social network, is more known and studied.

The use of models based on mobile phone data to study human activity first started with analysis of the calls graph. Indeed, studying how people are interconnected reveals their friendship community [6] and can be used to predict the age and gender of the user [7]. Other exemple include modeling the spectrum of ego patterns variability and revealing the social power (e.g. authority/popularity) of users [8, 9].

The analysis of mobile phone data also offers novel opportunities to researchers working on human mobility at a large scale. In the last ten years, many human mobility models were proposed. Generally, there are two main approaches to model human mobility: trip-based, where aggregated data are used [2, 9, 10] and activity-based when individual data are considered [2, 4, 11, 12]. Human mobility models were successfully applied in various topics like activities detection [11, 13], human mobility prediction [14, 15] or tourism applications [5, 12, 16].

Models based on mobile phone data are also used to measure urban dynamics and to partition space in communities [6, 17, 18]. Simulation models were also proposed. Morlot [19] analyzed phenomena related to user clumps and hot spots occurring in mobile networks at the occasion of large urban mass gatherings in large cities. The analysis is based on observations made on mobility traces of GSM users in several large cities. A mobility model is introduced and tested, based on interaction and mutual influence between individuals.

3 Acquisition of Mobile Phone Data

This section presents the primary methods that are used to collect data using mobile phone tracking technologies. Generally, there are two types of collection methods based on telecom networks: active collection and passive collection.

3.1 Active Collection

Active collection methods consist of collecting mobile phone location for a sample of participants who gave their agreements to be located. Three main type of methods allow active location:

1. **The Geolocation platform**. Each telecom operator has a geolocation platform which is currently used for advertisement purposes. In this technology the telecom network periodically sends a localized demand signal to the mobile phone. The location is obtained by translating the mobile phone answer into a geographical position which corresponds to the position of the GSM base station. Note that data collection is done through the available APIs proposed by each telecom operator.

The geolocation platform has a constant recording rate. Another advantage of this technology is that the sample is the set of all users in a given geographic area, and not only customers of a specific operator. The disadvantages are that the location is less accurate and heterogeneous depending on the geography of the GSM network.

2. **Phone integrated probe software**. This method consists of installing software in each mobile phone to capture data from mobile phone's sensors. Thus, if the mobile phone has a GPS chipset, its location measurements are recorded and stored or sometimes uploaded if the mobile phone has internet access. The location can also be estimated considering the GSM base station location (this information is encoded in the GSM protocol) or even the identifier of the Wi-Fi signals captured in the neighborhood and confronted to the Google API which returns then a location. However, this requires developing a specific communication data collection and transmission protocol. The recording rate is not constant and depends on both mobile phone use (communication) and device displacements.

3. **Phone GPS activation**. This method is based on GPS (Global Positioning System) technology. It consists of activating the GPS receivers for the smart mobile devices in order to determine their location. The main advantage of the GPS technology is that the location has accurate positioning which can be from 5 to 20 m. The disadvantages of GPS technology are that only participants having a Smartphone can be selected; besides, the GPS may not obtain signal due to obstacle and the power consumption of the receivers may quickly deplete a mobile phone's battery within a day.

As aforementioned, active collection can employ the base station method, the satellite-based system for navigation or Wi-Fi protocol to locate mobiles phones. The main disadvantage of this data collection technique is that it can only be applied to the (limited) sample of participants who accept to participate to the study. On the other side, the quality of collected data can be improved both in terms of location and semantic information by asking the participants to validate or to change some recorded locations and to add some motivations about the moves, stops, mean of the transportation, etc.

3.2 Passive Collection

Each operator collects and stores for a given period customers' mobile phones activities for billing or for technical telecom network measurements purposes. This type of collection is named passive collection, since recordings are made automatically. There are three main types of mobile phone data collected using the passive collection:

1. **Call Detail Records data (CDR)**. These data are automatically collected by the telecom operator for billing needs. The communication activities (calls in/out, SMS in/out, internet connection) of each mobile phone are recorded as rows

(id - record type - timestamp - location of the GSM base station). Record type has four values: call in/out, SMS in/out. The main advantages of CDR data are the big mass of geolocated data for a long period of time and the relation between users which can be used for graph analysis. The disadvantage comes from the high variability of human communications, since only the calls/SMS that occurred are part of the data.

2. **Probes data**. Considering the GSM network quality needs, another kind of row data are collected via a Mobile Switching Center (MSC) that maintains a database of signaling events concerning mobile phones present on its territory by tracking the geolocation activity. This technology can provide mobile phone locations when a communication activity occurs, as in the previous case, but also for activities such as switching on/off, location update after at least three hours of mobile phone inactivity, and itinerancy activity (i.e. records generated by a real mobile phone movement such as handover and LAC update). The handover is the change of the antenna during the communication and the LAC update is the change of a LAC (set of antennas) due to the movement of a person. This means that, even if the mobile phone is not used, a record occurs if a user enters in a new LAC. These data are formatted as (id - record type - timestamp- location of base station -LAC id). The record type could have the following values: call in/out, SMS in/out, switch on/off, call/ SMS handover, LAC update.

Figure 1 shows the temporal distribution of mobile phone events per hour. Mobile phone data are comming from probes (see 3.2) in Paris Region, France during

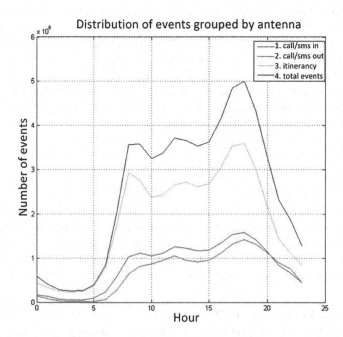

Fig. 1 Daily events distribution; the number of events is grouped by antennas and by hours

one week-day. As we can see, the majority of records are generated by itinerancy events (curve 3) instead of communication events (curves 1 and 2). Some peaks can be observed at 9, 12 am and 6 pm. These peaks are related to residency/work commuting and lunch periods coherent with the French daily activity rhythms.

Generally, the location of the two first types of data (CDR and probes) is limited to the GSM base station location. Nevertheless, the accuracy of location can be improved using the triangulation method (telecom operator and mobile phones can use signals from neighboring GSM base stations to estimate the location of the mobile phone) or by adding some information such as the received Signal Strength Indicator, the time difference of arrival and angle of arrival.

3. **Wi-Fi data**. A relatively new source of geolocation information is the use of Wi-Fi access points. This is possible since the Wi-Fi access points continuously notify their existence. The technology is quite similar to the use of GSM base station, i.e. the mobile phone connects to an access point which has a unique identifier and a geographic location. The passive collection consists of using an API such us Google maps, which has collected billions of couples (captured Wi-Fi signal id, base position). From the Wi-Fi carrier point of view it is also possible to localize each terminal connected to its hotspots.

This system is powerful in terms of precision and independence to GSM noise, but it requires two elements: first, the mobile phone has a data plan so that it can use an API via internet access, and second, an operator must provide this API of couples (Wi-Fi id, location of the base).

The passive collection allows recording big data for an important number of people and for a long period and provides an objective snapshot of user's behavior. Thus, considering that the penetration rate is around 120 % in Europe countries (see Table 1) and that historical carriers have 40 % of the market share, the mobile phone could be considered as a proxy of the inhabitants communication and location activity, which can be used to develop models of human communication and human mobility.

Table 2 presents a summary for the key technical differences between the six different technologies.

Furthermore, poor semantic information (e.g. the motivation of the move, the stop, the social and economic description of people), is available for these data, which makes analysis very complex from an application point of view. To improve this lack of information, many models were proposed in the literature. The first model that allows adding semantic information is based on the concept of trajectory and is proposed by [20]. Considering this model, a trajectory is denoted by a set of moves

Table 1 World wide mobile phone subscribers (per 100 people)

Developed nations	Developing nations	Africa	Arab States	Asia and Pacific	Europe
117.8 %	78.8 %	53.0 %	96.7 %	73.9 %	119.5 %

Source International Telecommunication Union (November 2011)

Table 2 Comparison of different tracking technologies

Type	Technology	Resolution	Sampling rate
Active	Geolocation	Depends on the cell size: medium in urban, low in rural	Constant
	Phone integrated probe software	Depends on the cell size: medium in urban, low in rural	Not constant
	Phone GPS activation	High	Constant
Passive	CDR	Depends on the cell size: medium in urban, low in rural	Not constant, depends on users's communication
	Probes	Depends on the cell size: medium in urban, low in rural	Not constant, depends on users's communication and move
	Wi-Fi data	Depends on the hotspot size: medium in urban, low in rural	Not constant, depends on users's Internet usage

and stops. The stops are places where people stop during an interval in order to do an activity. Moves are considering as continuous path in space and time between two consecutives stops.

Different methods were proposed to instantiate the model of stops and moves and especially to automatically detect stops in trajectories. Stops detection methods use different criteria such as speed criteria [21, 22], distances criteria [21, 27], speed and duration criteria [21, 24], spatial voronoi positions and time intervals [5]. Once stops and moves are automatically detected, there are models that propose to estimate the type of activity such as tourist activities [5, 25], cultural evenings [26], work and home [13, 23] or the mean of transportation between two consecutive stops [28, 29].

4 Data Integration and Uncertainty Management

Data coming from sensors such as mobile phones have uncertainties, imperfections and incompleteness depending on the acquisition mode. This is why the analysis of spatial mobility requires the use of a various data sources, from spatial data (e.g. roads, rail network, land use) to socioeconomic, demographic and historic data. It is essential to cross sensor data with the 'traditional' data in order to capture the complexity of human mobility and human behavior.

Spatial data sources are certainly growing. They are nonetheless increasingly heterogeneous and of varying quality. Moreover, in order to use them in decision-making processes (planning, development, epidemiology, and crisis) data quality in terms of accuracy must also be taken into account and modeled.

To our knowledge, few works take into account the imperfection of mobile data. Our goal is to define a typology of imperfection and to propose a model adapted for all. Four types of imperfection are distinguished: incompleteness, imprecision, uncertainty and spatiotemporal granularity.

- Incompleteness: refers to the lack of information. In our case, it concerns the data collected (LAC identifier, type of the event are not recorded) and the tracking process itself (due to a malfunction, events could not be recorded).
- Imprecision: is the difficulty to express clearly and precisely the reality. For example, the location of a person through his mobile phone is only defined at the scale of a cell.
- Uncertainty: expresses a doubt on the validity of information. For example, no attribute information (e.g. user' identifier, LAC identifier or cell identifier) exists for a given record.
- Spatial and temporal granularity. The concept of granularity, as is used here, concern spatial and temporal dimensions in the process of data collection. First, temporal granularity applies to the frequency of records, i.e. the delay between two consecutive records. This frequency is random for passive data depending both on the communication activities (call, SMS, etc.) and users move (handover and LAC update). Second, the spatial dimension of LAC is heterogeneous. In the urban level, it can vary between 3.5 and 8,000 km^2 (see Fig. 2). Finally, the spatial granularity also applies to the location records and corresponds to the location of antenna to which the mobile phone is connected. The location of the mobile phone depends on the coverage of antenna which is modeled by a Voronoi cell.

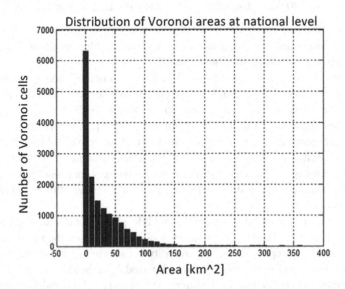

Fig. 2 Distribution of Voronoi areas at national level; each category corresponds to a bin having a crisp width (interval equals to 10 km^2) and an height equals to the number of Voronois belonging to this class

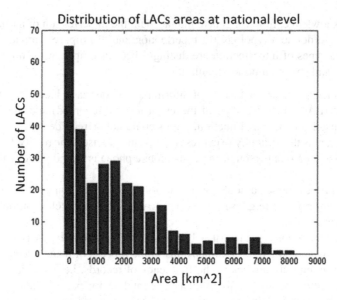

Fig. 3 Distribution of LACs areas at national level; each category corresponds to a bin having a crisp width (interval equals to 400 km^2) and an height equals to the number of LACs belonging to this class

Thus, the accuracy of the location of the mobiles phones via the telecom network is closely linked to the network structure, the Voronoi varying from and can vary from 3 m^2 to 300 km^2 depending on whether the user is located in a urban or rural environment (see Fig. 3).

Figures 2 and 3 are computed using one operator's telecom network structure (antennas and Lacs) in Paris Region, France.

Furthermore, the spatial granularity of location is not homogeneous: the antennas grid scalability depends, for example on the inhabitant's density, which is high inside urban areas and low in suburban or rural zones. Moreover the temporal resolution of recordings depends mainly on mobile phone usages frequency. These features produce heterogeneous information on individual and collective mobility. In order to handle these limitations, we propose to measure urban dynamics and to quantify the uncertainty of measured information. Thus, simple indexes are proposed to describe the mobility state (speed of displacement) and how confident this information is (uncertainty).

In order to build a mobility model via mobile phone data, a simple model is proposed to represent the sequential mobility of users based on their mobile phone tracks. The goal is to represent the user mobility but also to assess the quality of this model. As we show in Fig. 4, mobility can be inferred from the average speed (named theta and noted θ) between two sequential records, n and n − 1, considering laps-time $\Delta t = t(n) - t(n\text{-}1)$ and distance between the location of two sequential records far from each other from d = Dist(n − 1, n). θ is computed in Eq. 1.

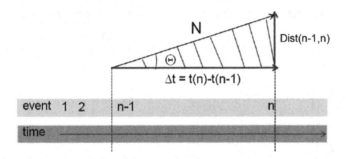

Fig. 4 Vectorial approach

$$\theta = a \, \tan \frac{d}{\Delta t} \tag{1}$$

The average speed between two records can also be expressed in a trigonometric way, so that the value can be between 0 and 1. It is noted V and computed in Eq. 2.

$$V = 1 - a \, \tan \frac{d}{\Delta t} \tag{2}$$

θ is a speed index describing the user mobility between two records during a temporal delay.

The second index assesses whether θ measurement is confident or if this value has only a mathematical ground. Thus, the uncertainty reflects how confident the measured mobility state is. Uncertainty is related to the Kullback Leibler divergence estimation [30] and is expressed as follow:

$$N = d \times \log \left(1 + \frac{d}{\Delta t}\right) + \Delta t \times \log \left(1 + \frac{\Delta t}{d}\right) \tag{3}$$

The greater the delay Δt or the distance d between two records is, the lower the confidence in the dynamic (speed variation or location transition) during the interval is. Moreover, the uncertainty is higher when one of the variables increases, even if the other one is stable (for example a long distance travelled in a short time or a short distance in a long time).

From the speed estimation and its confidence, we can derive the probable successions of position of the user during the time between two records; the quality of the user mobility increases with sampling frequency. Generally, two kinds of data are used to estimate user location: the communication ones (i.e. mobile phone data generated by the user during his calls/sms activity and the mobility ones (i.e. data generated by the telecom' technical network to localize mobile phone: LAC update, handover and location update after at least three hours of mobile phone inactivity). In order to obtain a sample of well defined trajectories, it is tempting to use only users having at least n communication records during a given day, so that

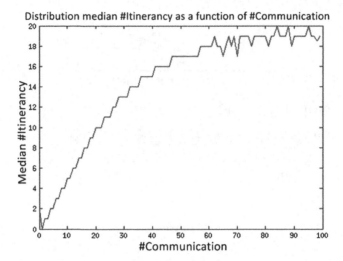

Fig. 5 Distribution of itinerancy as a function of communication

the associated frequency measurement is sufficient. But the estimation of correlation coefficient between the number of communication events and mobility events for a given user shows a significant correlation (correlation coefficient is equals to 0.81); it is confirmed via the Kolmogorov Smirnov pair test, where the hypothesis that the two distributions are different is rejected with a confidence greater than 0.95. The correlation between the itinerancy and communication events is shown in Fig. 5.

This result means that the more a user moves during a day the more likely to produce communications tracks, and reciprocally, the more a user calls during a day, the more he will produces mobility traces. Thus, while it is very tempting to increase the quality of mobility models by prefiltering individuals having a good communication sampling frequency, this result shows that a bias is introduced in the data sample, because the mobility model (properties) dependents on calls frequency.

5 Conclusion

In this chapter, the main mobile phone tracking technologies are first presented. We focus then on models that were proposed in order to study human mobility and behavior using mobile phone data. Finally, the question of uncertainty and data fusion is raised and we proposed both a typology of mobile phones imprecision and two metrics that allow to measure uncertainty.

Another problem which rises during the use of mobile phone data to model behaviors is the privacy. The question of privacy has been widely debated during the last decade. However, many solutions were proposed to protect users' privacy such as anonymisation, aggregation, intentional added uncertainty [31] and finite state machine to mask privacy [32]. From our experience and depending on the legislation of each country, privacy should be treated case by case according to the application.

Despite these imperfections, mobile phone data provide rich and massive information, obtained at low cost for a large spatial area (country scale) and over long time periods. The models based on these data can provide a richer understanding of human behavior patterns in space and time. Many questions referring to the use of mobile phone data for studying mobility still need to be explored: How to deal with the such a volume of data? Which filter applies to select people? Which scale should be used: individual, aggregate or multiscale? How to integrate contextual data such as social, economic, demographic data to improve the quality of semantic mobile phone data?

References

1. Ahas, R.: Mobile positioning in mobility studies. In: Büscher, M., Urry, J., Witchger, K. (eds.) Mobile Methods. Routledge. London (2010)
2. Gonzalez, M.C, Hidalgo, C.A, Barabasi, A.L.: Understanding individual human mobility patterns. Nature, **453**, 779–782 (2008)
3. Becker, R.A., Caceres, R., Hanson, K., Loh J.M., Urbanek, S., Varshavsky, J., Volinsky, C.: A tale of one city: using cellular network data for urban planning. In: Proceedings of IEEE Pervasive Computing (2011)
4. Reades, J., Calabrese, F., Sevtsuk, A., Ratti, C.: Cellular census: explorations in urban data collection. IEEE Pervasive Comput. **6**, 30–38 (2007)
5. Olteanu Raimond, A.M., Trasarti, R., Couronne, T., Giannotti, F., Nanni, M., Smoreda, Z., Ziemlicki, C.: GSM data analysis for tourism application. In: Proceedings of 7th International Symposium on Spatial Accuracy Assessment in Natural Resources and Envi-ronmental Sciences (2011)
6. Blondel, V., Guillaume, J.L., Lambiotte, R., Lefebvre, E.: Fast unfolding of communities in large networks. J. Stat. Mech. Theory Exp. **2008**, P10008 (2008)
7. Licoppe, C., Diminescu, D., Smoreda, Z., Ziemlicki, C.: Using mobile phone geolocalisation for socio-geographical analysis of coordination, urban mobilities, and social integration patterns. Tijdschrift voor Economische en Sociale Geografie **99**, 584–601 (2008)
8. Stoica, A., Prieur, C.: Structure of neighborhoods in a large social network. In: Proceedings of IEEE International Conference on Social, Computing (2009)
9. Couronné, T., Stoica, A., Beuscart, J.S.: Online social network popularity evolution: an Additive Mixture Model. In: Proceedings of International Conference on Advances in Social Networks Analysis and Mining (2010)
10. Sevtsuk, A., Ratti, C.: Does urban mobility have a daily routine? Learning from the Aggregate Data of Mobile Networks. J. Urb. Tech. **17**, 41–60 (2010)
11. Olteanu Raimond, A.M., Couronné, T., Fen-Chong, J., Smoreda, Z.: Le Paris des visiteurs, qu'en disent les téléphones mobiles ? Inférence des pratiques spatiales et fréquentations des sites touristiques en Ile-de-France. Revue Internationale de la Géomantique (to appear in septembre), (2012)
12. Ahas, R., Aasa, A., Roose, A., Mark, Ü., Silm, S.: Evaluating passive mobile positioning data for tourism surveys. An Estonian case study. Tourism Manag. **29**, 469–486 (2008)
13. Phithakkitnukoon, S., Horanont, T., Di Lorenzo, G., Shibasaki, R., Ratti, C.: Activity-aware map:identifying human daily activity pattern using mobile phone data. In: Proceedings of International Conference on Pattern Recognition, Workshop on Human Behavior Understanding, pp. 14–25. Springer, Heidelberg (2010)

14. Calabrese, F., Di Lorenzo, G., Ratti, C.: Human mobility prediction based on individual and collective geographical preferences. In: Proceedings of 13th International IEEE Conference on Intelligent Transportation Systems (2010)
15. Song, C., Qu, Z., Blumm, N., Barabasi, A.L.: Limits of predictability in human mobility, Sci. 327, 1018–1021 (2010)
16. Asakura, Y., Takamasa, I.: Analysis of tourist behavior based on the tracking data collected using a mobile communication instrument. Transp. Res. A 41, 684–690 (2007)
17. Blondel, V., Deville, P., Morlot, F., Smoreda, Z., Van Dooren, P., Ziemlicki, C.: Voice on the border: do cellphones redraw the maps?. Paris Tech. Rev. 15, http://www.paristechreview.com/2011/11/15/voice-border-cellphones-redraw-maps/ (2011)
18. Ratti, C., Sobolevsky, S., Calabrese, F., Andris, C., Reades, J., Martino, M., Claxton, R., Strogatz, S.H.: Redrawing the map of Great Britain from a network of human interactions. PLoS ONE 5, e14248 (2010)
19. Morlot, F., Elayoubi, S.E., Baccelli, F.: An interaction-based mobility model for dynamic hot spot analysis. In: Proceedings of IEEE Infocom (2010)
20. Spaccapietra, S., Parent, C., Damiani, M.L., De Macedo, J.A., Porto, F., Vangenot, C.: A conceptual view on trajectories. Data Knowl. Eng. 65, 126–146 (2008)
21. Palma, A.T., Bogorny, V., Kuijpers, B., Alvares, A.O.: A Clustering-based approach for discovering interesting places in trajectories. In: Proceedings of ACMSAC ACM Press, New York (2008)
22. Yan, Z., Parent, C., Spaccapietra, S., Chakraborty, D.: A hybrid model and computing platform for spatio-semantic trajectories. In: ESWC, The semantic web: Research and Applications, 7th Extended Semantic Web Conference, Heraklion, Greece, Springer Heidelberg, pp. 60–75 (2010)
23. Andrienko, G., Andrienko, N., Hurter, C., Rinzivillo, S., Wrobel, S.: From movement tracks through events to places: extracting and characterizing signicant places from mobility data. In: Proceedings of IEEE Visual Analytics, Science and Technology, pp. 161–170 (2011)
24. Zimmermann, M., Kirste, T., Spiliopoulou, M.: Finding stops in error-prone trajectories of moving objects with time-based clustering, intelligent interactive assistance and mobile multimedia. Computing. 53, 275–286 (2009)
25. Spinsanti, L., Celli, F., Renso, C.: Where you stop is who you are: understanding peo-ples activities, In: Proceedings of 5th BMI, Workshop on Behavior Monitoring and In-terpretation, pp. 38–52 Germany (2010)
26. Calabrese, F., Pereira, F.C., Lorenzo, G.D., Liu, L.: The geography of taste: analyzing cellphone mobility and social events. In: Proceedings of IEEE International Conference on Pervasive Computting (2010)
27. Andrienko, G., Andrienko, N., Olteanu Raimond, A.M., Symanzik, J., Ziemlicki, C.: Towards extracting semantics from movement data by visual analytics approaches. In: Proceedings of GIScience Workshop on GeoVisual Analytics, Time to Focus on Time in Columbus OH, to appear (2012)
28. Smoreda, Z., Olteanu Raimond, A.M., Couronne, T.: Spatio-temporal data from mobile phones for personal mobility assessment. In: Proceedings of 9th International Conference on Transport Survey Methods: Scoping the Future while Staying on Track, Termas de Puyehue, Chili (2011)
29. Wang, H., Calabrese, F., Di Lorenzo, G., Ratti, C.: Transportation mode inference from anonymized and aggregated mobile phone call detail records. In: Proceedings of 13th IEEE Conference Intelligent Transportation Systems, pp. 318–323 (2010)
30. Kullback, S., Leibler, R.A.: On information and sufficiency. Ann. Math. Stat. 22(1), 79–86 (1951)
31. Duckham, M., Kulik, L., Birtley, A.: A spatiotemporal model of strategies and counter-strategies for location privacy protection. In: Proceedings of the Fourth International Conference on Geographic Information Science. Schloss Münster, Germany (2006)
32. Reades, J.: People, places and privacy. In: Proceedings of International Workshop Social Positioning Method, Estonia (2008)

Chapter 5
Change Detection in Dynamic Political Networks: The Case of Sudan

Laurent Tambayong

Abstract Social Network Change Detection (SNCD) algorithm is applied to detect abrupt change in political networks of Sudan. In particular, SNCD compares the longitudinal normalized network-level measures against a predetermined threshold value. These measures are density, average closeness, and average betweenness over different time periods. Data extracted from *Sudan Tribune* in 2003–2008 is presented in a form of yearly two-mode networks of agents and organizations. The result shows that SNCD detects abrupt changes that correspond to actual historical events. In particular, the analysis shows that the foreign-brokered signings of multiple peace agreements served as a political solidification point for political actors of Sudan. This was a catalyst that caused three leaders to have emerged and lead the more compartmentalized yet faction-cohesive political networks of Sudan.

1 Introduction

The state of Sudan has undergone major domestic conflicts resulting in changes in its political map. Since 2003, the violent domestic conflict in Sudan has caused an estimated of 300,000 casualties and 2.7 million refugees [4]. This prolonged, escalated and, complex crisis is political in nature as there are many stakeholders involved at both agent and organization level. This glue to the social fabric of Sudan includes different tribes and semi-autonomous communities loyal to local leaderships [20]. Masked as an allegation of oppression of native Sudanese in favor of Arab Sudanese, the conflict has been centered in the issues of land and grazing rights [19]. The conflict has become worse when the government of Sudan allegedly encourages militias as a self-defense measure [4], creating social chaos. The formation of the militia results in further civil war-like conflict. The Sudanese government, although

L. Tambayong (✉)
California State University at Fullerton, Fullerton, USA
e-mail: ltambayong@fullerton.edu

V. Dabbaghian and V. K. Mago (eds.), *Theories and Simulations of Complex Social Systems*, Intelligent Systems Reference Library 52, DOI: 10.1007/978-3-642-39149-1_5, © Springer-Verlag Berlin Heidelberg 2014

showing characteristics of a dictatorship government, has been lacking in authoritative power. Thus, in desperation, the government has used an extreme—and most likely violent—measure in an attempt to control the endemically weak state. Consequently, this extreme measure resulted in the allegations of war crimes and genocide for some of the most important political figures of Sudan [7] such that they have attracted international attention and involvements. Given the violent struggles and significant foreign-influenced historical events in Sudan, the interest here is to apply a quantitative model capable of detecting abrupt changes in the political networks of Sudan and to analyze if there has been a rapid change in the structure of the political networks of Sudan which in turn has caused such violence.

In network science, there is a focus on micro-macro relationships where the longitudinal progression of networks in time is studied [11, 12, 16, 23, 27, 33, 34]. This line of research is referred by Wasserman, et al. [28] as the Holy Grail of social networks research due to its potentially practical and powerful implications. While many of the aforementioned approaches focus on prediction, McCulloh and Carley [21] argue that the focus should be on detection of change. They argue that, unlike prediction, detection aims to pragmatically identify change without the necessity of knowing its cause. This eliminates the problems of identifying the causality link of change: change in networks does not always correspond to its intrinsic evolution as it can also be affected by external forces such as exogenous sociopolitical shock [27, 34, 33]. This approach serves well for stakeholders who prefer a reactive approach as opposed to those who prefer an assumptions-laden proactive approach.

For this reason, Social Network Change Detection (SNCD) algorithm [21] is used in this paper to detect abrupt changes in the political networks of Sudan. SNCD compares the over-time [10, 12, 13, 17, 29, 31] normalized network-level measures of interests [6] against a predetermined threshold value (in the statistical sense of passing the threshold into the rejection region). When a measure passes the threshold, a change is identified as significant. For Social Networks Analysis, commonly-used measures are density [8], average closeness [14], and average betweenness [13]. In this paper, SNCD is applied against these three measures over different time periods to detect abrupt changes in the political networks of Sudan.

2 Data

The data in this paper represents the political networks of Sudan. It consists of the political actors of Sudan which are both agents (people) and organizations. The data is obtained from publicly available information: online newspaper. It consists of tens of thousands of newspaper articles from *Sudan Tribune* online in year 2003–2008. For each year, there are 2000–10000 articles extracted using the approach of Network Text Analysis (NTA). NTA is a branch of text-mining that encodes the text as a network of concepts and relationships [22], reflecting social and organizational relationships and structure [24]. The specific methodology used is grounded in established methodologies for identifying concepts, indexing the relations of words, syntactic grouping of words, and the hierarchical/non-hierarchical linking of words [18]. For the data in this paper, the extraction is proximity based using a window of seven: two actors are linked if they are within 7 concepts. Concepts are identified from a sentence after a thorough clean-up process. For example, consider this sentence: "President Omar Hassan Al Bashir and Ali Osman Mohammed Taha take a photograph together with Salva Kiir Mayardit." There are 18 words in this sentence. However, after a thorough clean-up process eliminating irrelevant common words without any specific meaning ("and," "take," "a," "together," and "with"), there are only 4 identified concepts in this sentence: "President Omar Hassan Al Bashir," "Ali Osman Mohammed Taha," "photograph," and "Salva Kiir Mayardit." For this paper, only concepts that are aliases of political actors are used. Thus, the concept "photograph" is not used while the concepts "President Omar Hassan Al Bashir," "Salva Kiir Mayardit," and "Ali Osman Mohammed Taha" are linked together as they are within 7 concepts to one another in a sentence. The detailed methodology of this NTA processes is presented in papers by Diesner, et. al. [9] and Tambayong and Carley [26].

The results of this extraction are political actors, agents and organizations, and their relationships for each year. In particular, they are represented as yearly directed two-mode networks of agents and organizations. These political actors include agents and organizations that have been reported to engage in some political, diplomatic, peace treaties, military, commerce, non-profit, and terrorism activities. For agents, they include alleged terrorists as Sudan has been designated as a state that sponsors terrorism by the U.S. State Department [35] as terrorism is politically motivated. These include social, political, and commercial organizations. Commercial organizations include Sudanese companies listed on the U.S. Office of Foreign Asset Control's SDN list for the same reason with the inclusion of agents allegedly connected with terrorism. Tribes of Sudan are also included in the organization category since tribe affiliation as social organization is very important in the social fabric of Sudan.

This data extraction is then organized in two sets of yearly two-mode networks: one set includes Sudanese-only agents and organizations while the other set also includes foreign agents and organizations with vested interests in Sudan. The number of Sudanese-only agent in each year is 28, 38, 33, 39, 41, and 39. The number of Sudanese-only organization in each year is 53, 72, 62, 74, 80, and 79. Thus, the sums of the total degrees of Sudanese-only two-mode networks are 81, 110, 95, 113, 121, and 118. When foreign actors are included, these numbers increase. The number of foreigner-included agent in each year becomes 139, 238, 209, 218, 214, and 223. The number of foreigner-included organization in each year becomes 195, 261, 254, 262, 293, and 311. Thus, the sums of numbers become 334, 499, 463, 480, 507, and 534.

3 Social Network Change Detection Algorithm for Statistical Analysis

Following McCulloh and Carley [21], SNCD works as follows.[1] An abrupt change in network is detected when mean of the measures μ_0 increases significantly such that for time t:

$$\sum_{i=1}^{t} X_i + t \left(\frac{\mu_0 + \mu_1}{2} \right) > A',$$

where the threshold $A' = \left(\frac{\sigma^2}{\mu_1 - \mu_0} \right) \log A$. It means that, setting the alternative hypothesis as $\mu_1 = \mu_0 + \delta \sigma_{\bar{x}}$, the test passes threshold into the rejection region when

$$\sum_{i=1}^{t} x_i + t \left(\frac{\mu_0 + (\mu_0 + \delta)}{2} \right) = \sum_{i=1}^{t} \left(x_i - \mu_0 - \frac{\delta}{2} \right) > A',$$

The standard deviation, δ, is calculated from the different time periods. Using the cumulative statistic, this decision rule Cumulative Sum (CUSUM) statistics C_t for time t becomes

$$C_t = \sum_{i=1}^{t} (Z_i - K) > A'$$

where $Z_i = \frac{\bar{x}_i - \mu_0}{\sigma_x}$ and $K = \frac{\delta}{2}$. Consequently, it follows that there are two CUSUM values used where the value $\pm h$ is predetermined threshold constants set based on the user's tolerance for error:

$$C_t^+ = \max\{0, Z_t - K - C_{t-1}^+\} > h^+$$
$$C_t^- = \max\{0, Z_t - K - C_{t-1}^+\} > h^-.$$

The positive CUSUM statistics corresponds to $\delta > 0$ when there is a significant *increase* and the negative one corresponds to $\delta < 0$ when there is a significant *decrease* of the measure used. If $C_t^{\pm}$ fluctuates around zero, then the network-level measure of interest is defined as functionally stable due to the lack of significant change. For the application of SNCD in term of CUSUM in this paper, the social networks measures used are density, average closeness, and average betweenness.

4 Results

The parameters on this analysis are set as follows. The value of threshold $\pm h$ is set at $h = \pm 3.5$. This value of h corresponds to $\alpha = 0.01$, meaning that the probability of the Type I error or false positive is at 1 %. The value of K is set at $K = 0.5$.

---

[1] For a detailed derivation of SNCD method, readers should consult McCulloh and Carley's [21] paper.

This corresponds to one standard deviation shift in the measure as $K = \frac{\delta}{2}$. The number of networks used for control behavior for μ_0 and σ is set at 2 (an average of the values from two time periods). The results of SNCD are presented for the CUSUM statistics given three network-level measures: average betweenness, average closeness, and density.

4.1 Sudan-Only Agents and Organizations

The SNCD analysis on Sudanese-only agents and organizations allows us to focus on the core of the political networks of Sudan. Focusing on the core allows us to study the intrinsic relationships of Sudan indigenous political networks.

The CUSUM chart for average betweenness centrality (Fig. 1) shows that there was a significant decrease that abruptly began in 2004 and passed the threshold in year 2006. The same figure also shows that there was a significant increase that abruptly began in 2007 then passed the threshold and peaked in 2008. Table 1 shows these abrupt and significant changes of this measure in year 2005 and 2008 (once the threshold had been passed, the values of the next years were no longer calculated as abrupt change had already been detected).

The CUSUM chart for average closeness centrality (Fig. 2) shows that there was no abruptly significant decrease or increase that passed the threshold in year 2003–2008. The same figure also shows that there was a noticeable but insignificant increase

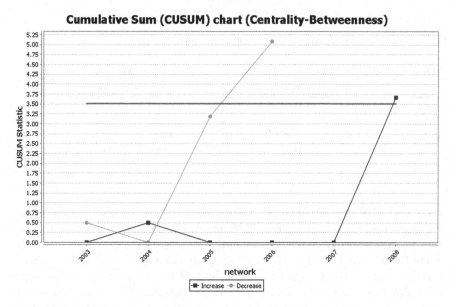

Fig. 1 CUSUM chart for average betweenness centrality (Sudanese only)

Table 1 CUSUM values for average betweenness centrality (Sudanese only)

	2003	2004	2005	2006	2007	2008
Value	0.01	0.01	0.01	0.01	0.01	0.01
Z-value	−1.00	1.00	−3.69	−2.40	−3.33	4.17
Increase	0.00	0.50	0.00	0.00	0.00	3.67
Decrease	0.50	0.00	3.19	5.08		

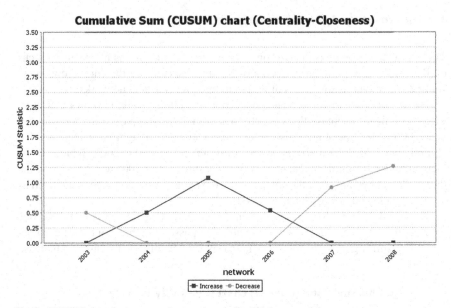

Fig. 2 CUSUM chart for average closeness centrality (Sudanese only)

Table 2 CUSUM values for average closeness centrality (Sudanese only)

	2003	2004	2005	2006	2007	2008
Value	0.00	0.00	0.00	0.00	0.00	0.00
Z-value	−1.00	1.00	1.08	−0.04	−1.42	−0.85
Increase	0.00	0.50	1.08	0.54	0.00	0.00
Decrease	0.50	0.00	0.00	0.00	0.92	1.27

that began in 2003 and peaked in 2005. The same figure also shows a noticeable but insignificant decrease that began in 2006 and peaked in 2008. Table 2 shows the relatively constant numbers for this measure.

The CUSUM chart for density (Fig. 3) shows that there was a significant increase that began and abruptly passed the threshold in 2007 then peaked in 2008. The same figure also shows that there was a noticeable but insignificant decrease peaking in

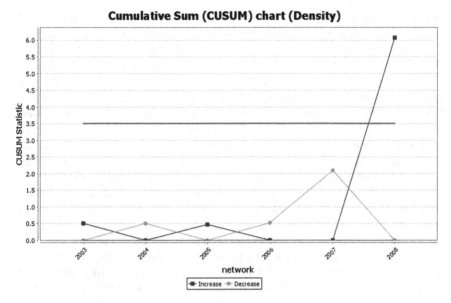

Fig. 3 CUSUM chart for density (Sudanese only)

Table 3   CUSUM values for density (Sudanese only)

	2003	2004	2005	2006	2007	2008
Value	0.02	0.02	0.02	0.02	0.02	0.03
Z-value	1.00	−1.00	0.98	−1.02	−2.07	6.58
Increase	0.50	0.00	0.48	0.00	0.00	6.08
Decrease	0.00	0.50	0.00	0.52	2.09	0.00

year 2007 that preceded the increase in 2008. Table 3 further shows that density peaked in year 2008 with a value of 0.03 and a Z-value of 6.58.

This CUSUM analysis can be used to explain the changes in power structure of the indigenous political networks of Sudan. However, brief reviews of political events in Sudan during the data collection period need to be presented first. Since 2003, the Darfur rebellion has ignited a series of important events that lead to a change in the landscape of Sudan's political networks. In 2004, the government of Sudan was accused of genocide by the international world following a series of violent and brutal incidents. Following the genocide accusation, the United Nations sanctioned the government of Sudan. This resulted in major changes in the political landscape of Sudan. In 2005, there was an international recognition of an autonomous government of South Sudan under John Garang, who also served as the First Vice President of Sudan replacing Ali Osman Mohammed Taha. Taha then became the Second Vice President of Sudan. Only a few months in office, Garang died in a helicopter crash. Garang's office as the First Vice President was then succeeded by

Salva Kiir Mayardit. Year 2005 also marked the beginning of a series of monumental signings of peace agreements. The first one was the Comprehensive Peace Agreement (CPA) signed in January 2005 by the Government of Sudan and the Sudan People's Liberation Movement/Army (SPLM/A). Following that, Darfur Peace Agreement was signed by the Government of Sudan and the Sudan Liberation Movement/Army (not to be confused with SPLM/A) in May 2006. Later, the Eastern Sudan Peace Agreement (ESPA) was signed in October 2006. ESPA ended comparatively lower-intensity conflicts in the eastern part of Sudan. However, violence seemed to have intensified again in the following year. In 2007, President Omar Hassan Ahmad Al-Bashir received war crime charges from the International Court. This resulted in an increased UN presence in 2008. These series of events are summarized in Table 4.

The CUSUM analysis can be now be further analyzed and related to these events. The CUSUM category of average betweenness (Fig. 1 and Table 1) detected an abrupt change, a decrease, which crossed the threshold in year 2005 and indicated that there was a sudden increase in compartmentalization of Sudan's political networks. This increase in compartmentalization corresponded to the changes in the vice presidencies of Sudan in 2005 as explained in the previous paragraph. These have been highly politicized positions serving as representations of power of various political organizations and regions. A complementary measure using the Eigenvector Centrality (Bonacich, 1987) helps to explain the effects of these vice presidency changes. Figure 4 shows the Eigenvector Centrality of the political agents of Sudan. For unfamiliar readers, Eigenvector Centrality reflects an agent's links to other well-linked agents. A higher score in this measure means that an agent is linked to well-linked other agents. Thus, this is a good measure for change in power structure. Figure 4 shows how Ali Osman Mohammed Taha's power declined in 2005 following his demotion from the First Vice President to Second Vice President of Sudan. It also shows the same trend for Garang who died in 2005 (Garang still showed some strength in this measures in the next couple years due to the amount of buzz in news published related to the controversies of his death). This figure also shows that Salva Kiir Mayardit's power increased sharply in 2005 following his promotion as the First Vice President. So, they remark how some agents gained while some other lost power. The agents whose eigenvector centrality measures notably increased such that they have occupied top ranks after the signings of the peace agreements are Mayardit and Al- Bashir. Although Taha's Eigenvector Centrality measure decreased in 2005, Fig. 4 shows that he stills ranked third in that year. The powerful measures of these three agents corresponded to the increase in compartmentalization of the political networks shown by the abrupt decrease in average betweenness: many other agents have chosen their political factions and have affiliated themselves with these powerful agents (Tables 5, 6 and 7).

The measures of power structure that have been discussed so far, eigenvector centrality and betweenness centrality, show that there have been three top agents in Sudan's political network since 2008: Al-Bashir, Taha, and Mayardit. Agreeing with results by Tambayong [27], this shows that the peace agreements solidified the position of these three figures as the most powerful political agent in Sudan. These results also show that year 2007–2008 also marked a significant change as indicated

Table 4 Summary of historical events in Sudan 2003–2008

	2003	2004	2005	2006	2007	2008
Event	Darfur rebellion	International recognition of Genocide	UN Sanctions Autonomous Southern Sudan CPA signed	DPA signed ESPA signed	War crime charges	Increased UN presence Sudan-Chad accord
Al-Bashir	President of Sudan	President of Sudan	President of Sudan	President of Sudan	President of Sudan	President of Sudan
Garang			Became VP of Sudan & Southern Sudan President. Died in Plane crash.			
Taha			Demoted to Second VP from First VP of Sudan			
Mayardit			Became First VP of Sudan			

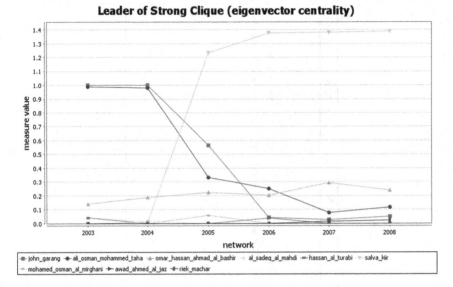

Fig. 4 Eigenvector centrality for top political agents of Sudan in 2003–2008 (Sudanese only)

Table 5 CUSUM values for average betweenness centrality (foreigners included)

	2003	2004	2005	2006	2007	2008
Value	0.01	0.01	0.01	0.01	0.01	0.01
Z-value	−1.00	1.00	−3.69	−2.40	−3.33	4.17
Increase	0.00	0.50	0.00	0.00	0.00	3.67
Decrease	0.50	0.00	3.19	5.08		

Table 6 CUSUM values for average closeness centrality (foreigners included)

	2003	2004	2005	2006	2007	2008
Value	0.00	0.00	0.00	0.00	0.00	0.00
Z-value	1.00	−1.00	4.15	9.34	−3.63	−10.83
Increase	0.50	0.00	3.65			
Decrease	0.00	0.50	0.00	0.00	3.13	13.46

by two CUSUM measures: average betweenness centrality and density. The CUSUM analysis shows that the values of average betweenness centrality (Fig. 1 and Table 1) and density (Fig. 3 and Table 3) increased abruptly such that they passed the threshold in 2007 and peaked in 2008. This means that the political networks of Sudan have become much more cohesive since 2008. Figures 5 and 6 show that Al-Bashir, Taha, and Mayardit occupied the top ranks of two other measures of network cohesion in 2003–2008: the Total Degree Centrality and the Clique Count [32]. This means that these three agents have become very strategically positioned in cohesive political

Table 7 CUSUM values for density (foreigners included)

	2003	2004	2005	2006	2007	2008
Value	0.01	0.01	0.01	0.01	0.01	0.01
Z-value	1.00	−1.00	−5.15	−5.55	−6.18	−1.10
Increase	0.50	0.00	0.00	0.00	0.00	0.00
Decrease	0.00	0.50	5.15			

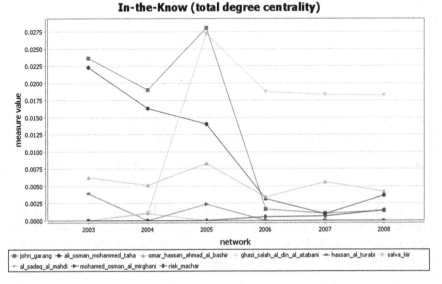

Fig. 5 Total degree centrality for top five political agents of Sudan in 2003–2008 (Sudanese only)

networks following the solidification point: the signings of the peace agreements. At this point, may I remind readers again that these are yearly two-mode networks of agents and organizations. So, the increase in cohesion was attributable not only to direct agent-agent relationships but also to indirect agent-organization-agent relationships. Accordingly, this power consolidation did not imply one unified power: it meant that these three leaders consolidated power to their own factions thus have created multiple consolidated factions.

4.2 The Influence of Foreign Agents and Organizations

This section presents SNCD analysis that includes foreign agents and organizations. Due to its violent nature, Sudan' conflict has drawn foreign involvements due to political and humanitarian concerns. Involved organizations include, but not limited to, the United Nations, U.S. government, the African Union, and Arab League.

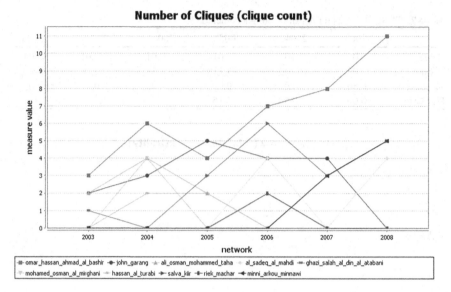

Fig. 6 Clique count for top five political agents of Sudan in 2003–2008 (Sudanese only)

Comparing the Sudanese-only (previous section) with the foreigner-included political networks allows us to assess the effects of foreign involvements in Sudan. This is important as foreign involvements might have catalyzed or inhibited changes in the political networks of Sudan by obscuring the intrinsic relationships of its indigenous networks.

In contrast to Sudanese-only result (Fig. 1), The CUSUM chart for average betweenness centrality does not show a significant increase or decrease that passed the threshold when foreigners were included (Fig. 7). However, there was still a significant and steady decrease that began in 2003 which became close to the threshold in year 2008.

On the other hand, while the Sudanese-only result did not pass the threshold (Fig. 2), the CUSUM chart for average closeness centrality shows both significant increase and decrease that passed the threshold when foreigners were included (Fig. 8). The increase that passed the threshold began in 2004 then both passed the threshold and peaked in 2005. The decrease began in 2006 then abruptly passed the threshold in 2007 and peaked in 2008.

Similar to the Sudanese-only result (Fig. 3), Fig. 9 shows that density passed the threshold when foreigners were included. However, while the Sudanese-only result shows that density increased and peaked in 2008, the foreigner-included result shows a significant decrease that began earlier in 2003 then abruptly passed the threshold in 2004 and peaked in 2005.

The above abrupt decrease of density and increase of average closeness suggest that foreign involvements caused the political networks of Sudan to reshuffle and compartmentalize around the solidification point (the signings of the peace

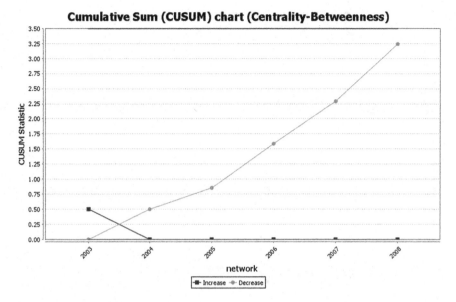

Fig. 7 CUSUM chart for average betweenness centrality (foreigners included)

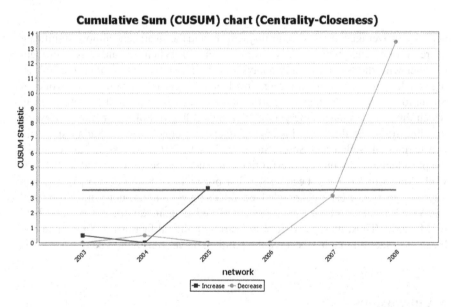

Fig. 8 CUSUM chart for average closeness centrality (foreigners included)

agreements). At this time, political actors seeked for suitable platforms and allies, foreigners included, that served their best interests. The signs of compartmentalization
as an effect of foreigner involvements were also shown by the abrupt and significant

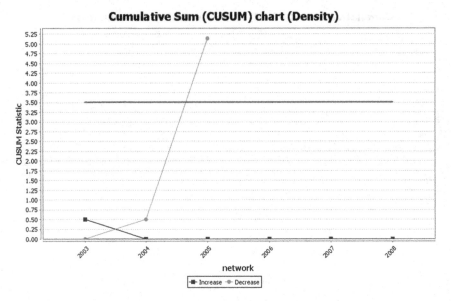

Fig. 9 CUSUM chart for density (foreigners included)

decrease in average closeness centrality (Fig. 8) and the steady and almost-significant decrease in average betweenness centrality (Fig. 7). When this analysis is brought together with the Sudanese-only analysis in the previous section, then the following conclusion can be drawn: foreign involvements caused the political networks of Sudan to first reshuffle then compartmentalize following the solidification point in which three leaders have emerged to lead their cohesive political factions. Beyond the time frame of this data, history validates this conclusion as Sudan split into two nations of Sudan and South Sudan in July 9, 2011: Al-Bashir has remained the president and the most powerful figure in Sudan as he got reelected in 2010 [5] while Mayardit has since become the first president of South Sudan [3]. Later in the same year, Al-Bashir acknowledged Taha's power and promoted him from Second Vice President to First Vice President of Sudan again, an office previously held by Mayardit. Thus, there have been two strong political factions in Sudan and one breakaway faction that has become the political power of South Sudan.

5 Conclusions

The results show that the CUSUM statistics of Social Network Change Detection algorithm is capable of identifying abrupt changes in the measures of average betweenness, average closeness, and density of the political networks of Sudan. Yet, although SNCD detects significant historical changes in Sudan, it is incapable of future prediction. The advantage of SNCP is also its weakness: it does not require the cause of the change to be known which also means that the cause of change is not identified, whether it is endogenous or exogenous. Nonetheless, SNCD analysis is useful to gain a deeper understanding of change and supports evidence based explanation. Using SNCD, the specific times when such abrupt changes occurred was identified and compared to the actual historical events of Sudan. For this purpose, the temporal progression of several other common Social Networks Analysis measures was also measured in helping to explain the causes of the abrupt changes. It followed that the foreign-brokered signings of multiple peace agreements served as a political solidification point that served as a catalyst for such abrupt changes. Following the solidification point, the results show that foreign involvements instigated the political networks of Sudan to reshuffle then compartmentalize such that three leaders, Al-Bashir, Taha, and Mayardit, have emerged to lead cohesive political factions. Their top ranks mean that they have been strategically positioned as powerful political agents in the more compartmentalized yet faction-cohesive post-solidification political networks of Sudan. History beyond the scope of this data also supports this analysis and validates the prominence of the three identified powerful political actors.

6 Discussions

The change detection of SNCD is sensitive to the values of the parameters: (i) the magnitude of change in SNCD statistics depends on the values of control behaviors (μ_0 and σ) and standard deviation shift (K) applied; (ii) the selection of the value of threshold (h) determines whether the change is defined significant or not (falling into rejection region or not). Thus, depending on the sensitivity of the data, the imprudent choice for the values of the parameters might yield to either type I or type II errors. This also brings up the issue that the normal distribution is implicitly assumed. In addition, SNCD analysis also only detects abrupt change and neither: (i) predict change; nor (ii) explain why change happens. Last but not least, SNCD method only detects change and does not describe the state of network. A paper by Tambayong [27] not only measures the stability of these networks but also explains the states they are in: growing, declining, or collapsing. All of these should be put into consideration in future methodology improvements of SNCD.

Acknowledgments This work is partly funded under a MURI—grant number N00014-08-1-1186 (Kathleen M. Carley, P.I.). The views and conclusions contained in this document are those of the authors and should not be interpreted as representing the official policies, either expressed or implied, of the Office of Naval Research, or the U.S. government. I am also grateful for generous advice and thoughts from Jeffrey C. Johnson and Richard A. Lobban.

References

1. BBC News: Darfur war crime suspect profiles. BBC News. http://news.bbc.co.uk/2/hi/africa/4946986.stm (2006). Accessed 20 July 2010
2. BBC News: Q&A: Sudan's Darfur crisis. BBC News. http://news.bbc.co.uk/2/hi/africa/3496731.stm (2009). Accessed 28 Jan 2010
3. BBC News: South Sudan's flag raised at independence ceremony. BBC News. http://www.bbc.co.uk/news/world-africa-14092375 (2011). Accessed 18 Sept 2012
4. BBC News: Sudan denies Darfur Militia ties. BBC News. http://news.bbc.co.uk/2/hi/africa/3908645.stm (2004). Accessed 28 Jan 2010
5. BBC WorldService: President Omar Al-Bashir re-elected in Sudan elections. BBC World Service. http://www.bbc.co.uk/worldservice/news/2010/04/100426_sudan_elections_hs.shtml (2010). Accessed 29 June 2010
6. Bonacich, P., Oliver, A., Snijders, T.A.B.: Controlling for size in centrality scores. Soc. Netw. **20**, 135–141 (1998)
7. CNN: Sudanese President charged with Genocide. CNN. http://www.cnn.com/2008/WORLD/africa/07/14/darfur.charges/ (2008). Accessed 28 Jan 2010
8. Coleman, T.F., Moré, J.J.: Estimation of sparse Jacobian matrices and graph coloring problems. SIAM J. Numer. Anal. **20**, 187–209 (1983)
9. Diesner, J., Carley, K.M., Tambayong, L.: Mapping socio-cultural networks of Sudan from open-source, large-scale text data. Comput. Math. Organ. Theory **18**(3), 328–339 (2012) (Special Issue: Data to Model)
10. Doreian, P.: On the evolution of group and network structures II: structures within structure. Soc. Netw. **8**, 33–64 (1983)
11. Doreian, P., Frans, N. (eds.): Evolution of Social Networks. Gordon and Breach, Netherland (1997)
12. Frank, O.: Statistical analysis of change in networks. Stat. Neerl. **45**, 283–293 (1991)
13. Freeman, L.C.: A set of measures of centrality based on betweenness. Sociometry **40**, 35–41 (1977)
14. Freeman, L.C.: Centrality in social networks I: conceptual clarification. Soc. Netw. **1**, 215–239 (1979)
15. Holland, P., Leinhardt, S.: A dynamic model for social networks. J. Math. Sociol. **5**, 5–20 (1977)
16. Huisman, M., Snijders, T.A.B.: Statistical analysis of longitudinal network data with changing composition. Sociolog. Methods Res. **32**, 253–287 (2003)
17. Katz, L., Proctor, C.H.: The configuration of interpersonal relations in a group as a time-dependent stochastic process. Psychometrika **24**, 317–327 (1959)
18. Kelle, U.: Theory building in qualitative research and computer programs for the management of textual data. Sociolog. Res. Online **2**, 2 (1997)
19. Malek, C.: The Darfur region of Sudan. http://www.beyondintractability.org/case_studies/Darfur.jsp?=5101 (2005). Accessed 29 Jan 2010
20. McCrummen, S.: A town constantly on brink of chaos. The Washington Post. http://www.washingtonpost.com/wp-dyn/content/article/2009/04/24/AR2009042403746.html (2009).Accessed 28 Jan 2010
21. McCulloh, I., Carley, K.M.: Detecting change in longitudinal social networks. J. Soc. Stuct. **12**, 3 (2011)
22. Popping, R.: Computer-Assisted Text Analysis. Sage, Thousand Oaks (2000)
23. Snijders, T.A.B.: Testing for change in a digraph at two time points. Soc. Netw. **12**, 539–573 (1990)
24. Sowa, J.F.: Concept Structures: Information Processing in Mind and Machine. Addison-Wesley, Reading (1984)
25. Sudan Tribune: Sudan's Bashir promotes Taha to first Vice-President and appoints a Darfurian as VP. Sudan Tribune. http://www.sudantribune.com/Sudan-s-Bashir-promotes-Taha-to,40146 (2011). Accessed 18 Sept 2012

26. Tambayong, L., Carley, K.M.: Political networks of Sudan: a two-mode dynamic network text analysis. J. Soc. Struct. **13**, 2 (2012)
27. Tambayong, L.: The stability and dynamics of political networks of Sudan. J. Artif. Soc. Soc. Simul. (2013) (forthcoming)
28. Wasserman, S., Scott, J., Carrington, P.: Introduction. In: Carrington, P., Scott, J., Wasserman, S. (eds.) Models and Methods in Social Network Analysis. Cambridge Press, Cambridge (2007)
29. Wasserman, S.: Stochastic Models for Directed Graphs. Ph.D. Dissertation, Harvard University, Department of Statistics, Cambridge, MA (1977)
30. Wasserman, S.: Analyzing social networks as stochastic processes. J. Am. Stat. Assoc. **75**, 280–294 (1980)
31. Wasserman, S., Iacobucci, D.: Sequential social network data. Psychometrika **53**, 261–282 (1988)
32. Wasserman, S., Faust, K.: Social Network Analysis: Methods and Applications. Cambridge University Press, New York (1994)
33. White, D.R., Tambayong, L., Kejžar, N.: Oscillatory dynamics of city-size distributions in world historical systems. In: Modelski, G., Devezas, T., Thompson, W. (eds.) Globalization as Evolutionary Process: Modeling, Simulating, and Forecasting Global Change. Routledge, London (2008)
34. White, D.R., Tambayong, L.: City system vulnerability and resilience: oscillatory dynamics of urban hierarchies. In: Dabbaghian, V., Mago, V. (eds.) Modelling and Simulation of Complex Social System. Springer, New York (2013)
35. U.S. Department of State: Sudan country reports on terrorism 2008, Chapter 3: State sponsors of terrorism. U.S. Department of State Office of the Coordinator for Counterterrorism. http://www.state.gov/s/ct/rls/crt/2008/122436.htm (2009). Accessed 16 Feb 2010

Chapter 6
High-Level Simulation Model of a Criminal Justice System

V. Dabbaghian, P. Jula, P. Borwein, E. Fowler, C. Giles, N. Richardson, A.R. Rutherford and A. van der Waall

Abstract Criminal justice systems are complex. They are composed of several major subsystems, including the police, courts, and corrections, which are in turn composed of many minor subsystems. Predicting the response of a criminal justice system to changes in subsystems is often difficult. Mathematical modeling can serve as a powerful tool for understanding and predicting the behavior of these systems under different scenarios. In this chapter, we provide the process flow of the criminal justice system of the British Columbia, Canada. We further develop a system dynamics model of the criminal justice system, and show how this model can assist strategic decision-makers and managers make better decisions.

1 Introduction

Common law criminal justice systems are composed of complex subsystems, such as police, courts, and corrections, operating in an interconnected yet autonomous fashion. These complex subsystems both contribute to, and respond to the operational fluctuations amongst their own elements, in other subsystems and across the system as a whole. Criminological analysis that focus on single components of a justice system are often unable to predict the effects of the changes in the policies or operations, on the other subsystems or divisions of the system (see [8, 12, 15]).

V. Dabbaghian (✉) · P. Borwein
MoCSSy Program, The IRMACS Centre, Simon Fraser University, Burnaby, BC, Canada
e-mail: vdabbagh@sfu.ca

P. Jula
Technology and Operations Management, Beedie School of Business, Simon Fraser
University, Burnaby, BC, Canada

E. Fowler · C. Giles · N. Richardson · A.R. Rutherford · A. van der Waall
Complex Systems Modelling Group, The IRMACS Centre, Simon Fraser University,
Burnaby, BC, Canada

V. Dabbaghian and V. K. Mago (eds.), *Theories and Simulations of Complex
Social Systems*, Intelligent Systems Reference Library 52,
DOI: 10.1007/978-3-642-39149-1_6, © Springer-Verlag Berlin Heidelberg 2014

This chapter is based on a recent application of systems thinking to the Criminal Justice System (CJS) in British Columbia, Canada [11], and its purpose is to illustrate how systems thinking can be used to assist the process of strategic decision-making in complex managerial situations.

The main goal of this article is to develop a general high-level system dynamics model of the CJS by analyzing and examining the effect of changes to the system as entities move through it over time. The system dynamics approach models entities as they move through the various stages of the CJS through a network of flow pipes. System dynamics is an ideal approach to modeling in such cases in comparison to other modeling approaches, because the data requirements are minimal and much of the behavior of the system can be constructed using expert knowledge of the system.

Developed in the early 1960s, system dynamics models are based on the economics concepts of stocks and flows. The strength of the system dynamics modelling lies largely on the fact that it is both qualitative and quantitative in nature. The qualitative nature of the system dynamics comes from the fact that it builds upon a systems thinking model approach. Thus the original model can be created using minimum data, by solely focusing on the qualitative nature of the system. This is best obtained based on the experience and the insights of professionals and managers. This knowledge base, although qualitative, is the foundation of the actual decision process in the system, and as such is considered more comprehensive.

The proposed system dynamics simulation model could be used to examine the impact of the management decisions on the performance of various components of a justice system, such as prosecution services, court services, and corrections. Furthermore, it could be used to determine the impact of other factors—both internal and external—on the performance of the various components of a criminal justice system. This model further addresses an important operational need for a simulation tool which can incorporate measures of workload. For example, how would a change in the number of court appearances will impact the efficiencies within the system? This model could be used to address questions such as:

- What is the impact of police resources on crime in general?
- How to allocate police resources in order to balance the workload across a criminal justice system?
- How does the distribution of police resources affect specific crime types, such as impaired driving or organized crime?
- What is the impact of changes in upstream practices on corrections — especially with regards to community versus institutional sentences?
- What is the interdependence between increases in pending cases and the workload for actors within a criminal justice system?
- What is the relationship between remand counts and average time to disposition for those in remand?

Furthermore, this model can help with understanding the principal drivers of resource consumption within a criminal justice system. It could be used to develop good metrics for resource consumption by answering questions such as: how to determine when resources are being used efficiently? how does resource consumption relate to dollar costs? and what is the best granularity with which to track resource consumption?

Many stakeholders will benefit from this model by running "what/if" scenarios. Policy makers, governmental agencies, police, justice department, and many private firms are among institutes that can use this model to make better decisions.

The rest of this article organized as follows: next section is an introduction to the criminal justice systems. In Sect. 3 the historical background of the research in the domain of modelling criminal justice systems are reviewed. Section 4 contains the description of the sub-systems in the proposed model, and the assumptions that are considered in this model. In Sect. 5 some simulation details such as the data and the verification and validation of the model are discussed. Section 6 contains examples of two scenarios that are studied using the proposed model. Finally, Sect. 7 covers the conclusions and the related future works.

2 Criminal Justice System

A criminal justice system (CJS) is a set of legal and social institutions for enforcing the criminal law in accordance with a defined set of procedural rules and limitations. CJSs include several major sub-systems, composed of one or more public institutions and their staffs: police and other law enforcement agencies; trial and appellate courts; probation and parole agencies; custodial institutions (jails, prisons, reformatories, halfway houses, etc.); and departments of corrections (responsible for some or all probation, parole, and custodial functions). Some jurisdictions also have a sentencing guidelines commission. Other important public and private actors in this system include: defendants; private defense attorneys; bail bondsmen; other private agencies providing assistance, supervision, or treatment of offenders; and victims and groups or officials representing or assisting them (e.g., crime victim compensation boards). In addition, there are numerous administrative agencies whose work include criminal law enforcement (e.g., driver and vehicle licensing bureaus; agencies dealing with natural resources and taxation).

The notion of a "system" suggests something highly rational and carefully planned, coordinated, and regulated. Although a certain amount of rationality does exist, much of the functioning of criminal justice agencies is unplanned, poorly coordinated, and unregulated. Most jurisdiction do not reform all (or even any substantial part) of its system of criminal justice. Existing systems may include some components that are old (e.g., jury trials) alongside others that are of quite recent origin (e.g., specialized drug courts). Moreover, each of the institutions and actors listed above has its own set of goals and priorities that sometimes conflict with those of other institutions and actors, or with the goals and priorities of the system as a whole. Furthermore, each of these actors may have substantial unregulated discretion in making particular decisions (e.g., the victim's decision to report a crime; police and prosecutorial discretion whether and how to apply the criminal law; judicial discretion in the setting of bail and the imposition of sentence; and correctional discretion as to parole release, parole or probation revocation, prison discipline, etc.).

Nevertheless, all of the institutions and actors in the CJS are highly interdependent. What each one does depends on what the others do, and a reform or other change in one part of the system can have major repercussions on other parts. It is therefore very useful to think about criminal justice as a system; not only to stress the need for more overall planning, coordination, and structured discretion, but also to appreciate the complex ways in which different parts of the system interact with each other.

3 Simulation Modeling in Criminal Justice Systems

Mathematical modeling and simulation of the CJS date back to 1970s. During that time there was a surge of activity using system dynamics model to represent CJSs (see [3, 5–7]). Some of the researches emphasized on modelling the court process [8, 12, 13]. In the recent years in parallel to the computer technology, mathematical modeling and simulation techniques in criminology and CJS are developed and expanded [1, 4, 9].

Previous studies on modelling the CJS have mainly focused on one of the subsystems in the CJS. Much of the simulation modeling focuses on police and its resources [16]; on the court system or components of the court system [8, 10, 14] or corrections, primarily prison systems [2, 17].

New interest in the use of system dynamics models in the CJS has blossomed among academics and professional in recent years for at least four separate, but related reasons. First, the continuing expansion of computing power has made it possible to construct and use large-scale simulation models that were simply not possible before. Second, the development of large electronic data archives have made model calibration and validation possible in ways that were not previously feasible. Third, the growing sophistication of criminal justice professionals has led them to ask researchers to explore policy impact questions of increasing complexity. Fourth, the software packages available today that can be used to construct system dynamics models were not readily available to researchers a couple of decades ago.

Next section describes the major components of the contemporary CJS in British Columbia. It presents a system dynamics model for better understanding of how these components typically operate in practice, and examines the various uses of the system concept.

4 Description of the System

In any modeling project the goals and assumptions of the model are inter-related. It is important to outline the assumed conditions and purpose of the model, as these affect the nature of the predictive "what-if" questions that a model is able to address, the level of model detail to consider and the data requirements of the model.

The first assumption of our model is that we are dealing with aggregate flows of entities over a unit of time. For instance, these entities can be crime incidents, reports to Crown Council, cases, persons, etc. System dynamics analyzes input data and examines the effect of changes to the system as these aggregate entities move through it over time. The system dynamics approach models entities as they move through the various stages (represented as *boxes*) of the CJS through a network of flow pipes (*arrows*).

The second assumption of this model is that it is a high-level model of a CJS. As a result, a) each of the parts of the system are modeled as stages where entities continue through the system or leave the system in various ways. This means that the full complexity at each stage might not be modeled, and b) this model is linear which means that *feedback behavior* is not modeled. An example of feedback behavior occurs when a stock (e.g., *Pending Cases* — see Sec. 4.3) has a capacity or limit. Once that limit is reached or exceeded the behavior of other system components or stocks changes over time.

Figure 1 shows the process flow of the CJS understudy. In this section we describe the details of each sub-system. The description of the model starts with the police sub-system, moves to the Crown and courts sub-system, continues with the corrections 1 process and concludes with corrections 2 sub-system. The details of these sub-systems are depicted in Figs. 2, 3, 4, 5, 6 and are explained in Sect. 4.1 to 4.4. The complete detailed flowchart can be found in Appendix A.

4.1 Police Sub-System

In the police sub-system, monthly inputs of Uniform Criminal Reports(UCR) reported offences enter the system as *founded offences* into the *Substantiated offences* stock. A certain proportion of the UCR offences leave the *Substantiated offences* stock as *not cleared* and the remaining offences are either *cleared by charge* or are *cleared by other means*. Currently, the *cleared by other means* flow is a system exit point. The monthly flow of offences *cleared by charge* is converted into the flow of

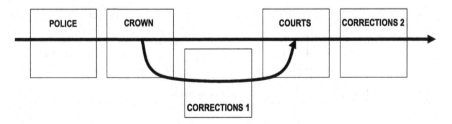

Fig. 1 Main sub-systems of the B.C. criminal justice system

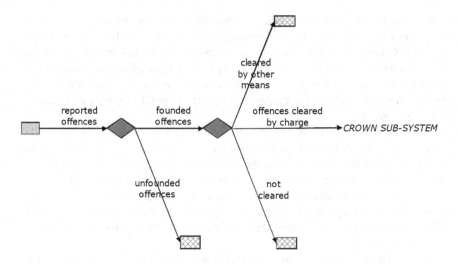

Fig. 2 Police sub-system of the B.C. criminal justice system

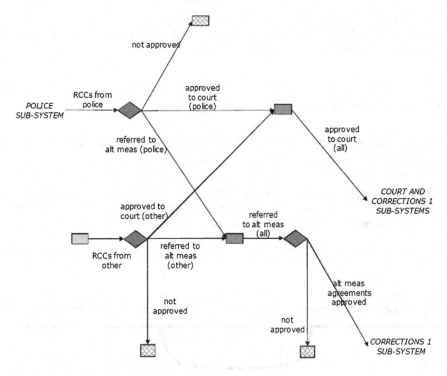

Fig. 3 Crown sub-system of the B.C. criminal justice system

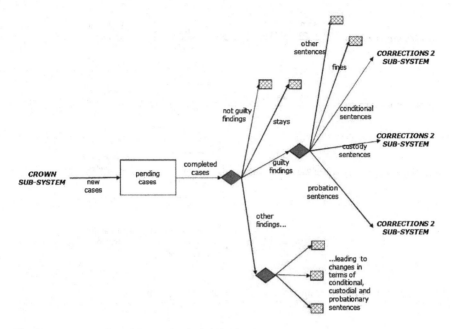

Fig. 4 Courts sub-system of the B.C. criminal justice system

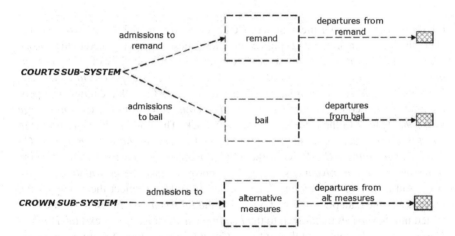

Fig. 5 Corrections 1 sub-system of the criminal justice system

Report to Crime Council from police *(RCCs from police).* Figure 2 shows the police sub-system.

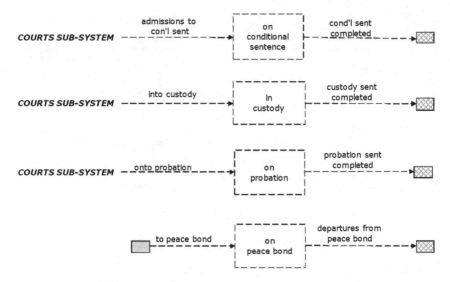

Fig. 6 Corrections 2 sub-system of the B.C. criminal justice system

4.2 Crown Sub-System

Flows of Report to Crime Council *RCCs* enter the model from two sources: (1) regulatory agencies, and (2) police agencies. These reports proceed to the *Crown Charge Assessment* stock. This stock determines the monthly flow of *RCCs not approved to court or sent to alternative measures* and *RCCs approved to court*. The Crown Charge Assessment stock provides a mechanism to track the number of people approved to court. The *RCCs approved to court* flow from the *Crown Charge Assessment* stock to the *Informations Sworn* stock. The flow of *RCCs approved to court* is used to determine the monthly values for *remand, bail and Released On Own Recognizance (ROOR)*, while the *RCCs not approved to court* dictates the flow of people into *alternative measures*. RCCs approved leave the *Informations Sworn* stock and flow into the *Pending Cases* stock. In order to reflect the reality of the process and the data provided RCCs leaving the *Informations Sworn* stock are converted into *new cases* that flow into the *Pending Cases* stock. Cases exit the *Pending Cases* stock each month and flow into the *Court Decision* stock. Figure 3 shows the Crown sub-system.

4.3 Court Sub-System

The *Court Decision* stock is not a separate stock in reality and merely represents a decision branch where cases may be resolved in one of four manners. They may be *Found Guilty* (either through a plea or trial), they may be *Found Not Guilty*, cases may be *Stayed*, or cases may proceed to an *Other* outcome. For the purposes of interpreting this model, the stocks of *Found Not Guilty*, *Stayed* and *Other* are system exit points. Cases in the *Found Guilty* stock flow into the *Sentencing* stock where they have one of five possible paths. The three sentencing flows that have an impact on the Corrections sub-system are: *Custody Sentences*, *Probation Sentences* and *Conditional Sentences*. The *All Fine Sentences* and *Other Sentences* flows are considered system exit points in the model. Figure 4 shows the court sub-system.

4.4 Correction Sub-Systems

The Correction subsystems have two main components, pre-conviction statuses and past-conviction statuses.

4.4.1 Corrections 1— Pre-Conviction Statuses

This sub-system in the justice model is the remand, bail or ROOR sub-system (termed the *Corrections Sub-system 1: Pre-conviction Statuses*) where accused not yet convicted of an offense, but detained by the police, have a status that precedes a court decision. Currently, this sub-system occurs in the flow of *RCCs approved to court* to the *Informations Sworn* stock before *new cases* enter the *Pending Cases* stock. This sub-system tracks the number of people approved to court in each of the three pre-conviction corrections stocks: *Remand*, *Released on Bail* and *ROOR* in each month. Each month a given number of people, determined through a formula, enter and leave each of these stocks. Additionally, the sub-system models the non-conviction based outcome *Alternative Measures*. This stock is derived from the flow of *RCCs not approved to court* leaving the *Crown Charge Assessment* stock. The *Alternative Measures* stock tracks the number of people on an alternative measures program and those that finish a program in each month.

4.4.2 Corrections 2 — Past-Conviction Statuses

The final sub-system is the corrections sub-system (referred to as the *Corrections Sub-system 2: Post-conviction Statuses*). This sub-system contains a non-conviction based outcome and conviction-based outcomes. The non-conviction outcome is a *Peace Bond* stock that has a separate input not linked to the rest of the model. This stock calculates the number of people under a peace bond and those that finish a peace bond each month. The conviction-based outcomes in the Corrections sub-system are composed of three stocks. The first stock is *In Custody* and it calculates the number of people in custody per month as a function of the number of people sentenced to custody and the number of people who have completed a custodial sentence. The second and the third ones are *On Probation* and *On Conditional Sentence* stocks which they operate in an identical fashion. Each of these stocks calculate the number of persons under each of these sentencing outcomes in a given month and, at present, do not account for resources.

5 Simulation Development

To develop the model, many meetings with different stakeholders have been organized. The initial step was to define the problem, in which the goals and the needs of the study was clarified. Next, the details of the components of the system was defined. We spent numerous hours understanding the actual behavior of the system, and mapping the process flow as described in the previous section. Determining the basic requirements of the model was necessary in developing the right model. Then, the required data was identified and collected. The model was translated to the simulation package. Our model has been subjected to rigorous verification and validation, in which we confirmed the model behave as intended, and ensured that no significant difference exists between the model and the real system (considering the assumptions).

We used the software package iThink[1] for simulation of the CJS model. This simulation contains the Police, Crown, Courts and Corrections subsystems as described in Sect. 4. It covers the flow of all activities, from the reporting of offences to the disposition of matters according to their many possible outcomes. For example, Fig. 7 presents the iThink model of the courts subsystem of the CJS as described in Section 4.3.

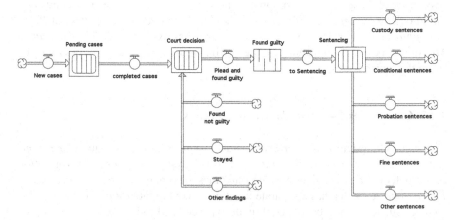

Fig. 7 iThink model of the courts sub-system of the B.C. criminal justice system

---

[1] http://www.iseesystems.com/

The model showed it is effective in producing quality results, and computationally efficient in-terms of the time required to generate the results. The simulation is done on a dual processor Apple G5 computer and the estimated computation time is about 30 s for each simulation run. In the following sub sections, we discuss examples of the efforts in gathering the data, and the verification and validation of the model.

5.1 Data

Once the agreed upon structure for the model was implemented, 5 years of aggregate system data for police, Crown, courts and corrections was used to populate the model, and to develop formulae for system behaviour. The data, provided by the Working Group of the Ministry of Public Safety and Solicitor General and the Ministry of Attorney General (PSSG/AG) in British Columbia, was a monthly count of entities in each stock from January 2003 to December 2007. The input data used in the model includes offences reported to police, RCCs from other agencies, and admissions to peace bond. The following is the list of the data fields that were used to construct and populate the model.

- **The Police Data Fields**

    1. UCR Data: Number of UCR Reported Offences.
    2. Clearance Data: Number of UCR Cleared by Charge and UCR Cleared By Other Means.

- **The Crown Data Fields**

    1. RCC Data: Crown Total RCCs, Crown RCCs Approved to Court, Crown RCCs Not Approved to Court.
    2. Person Data: Crown Total Persons, Crown Persons Approved to Court, Crown Persons Not Approved to Court.

- **The Court Data Fields**

    1. Court Case Data: Courts Provincial Criminal New Cases, Courts Provincial Criminal Completed Cases, Courts Provincial Criminal Pending Cases.
    2. Court Case Decision Data: Courts Provincial Criminal Guilty Findings, Courts Provincial Criminal Not Guilty Findings, Courts Provincial Criminal Other Findings, Courts Provincial Criminal Total Findings.
    3. Court Sentencing Data: Courts Provincial Criminal Custody Sentences, Courts Provincial Criminal Probation Sentences, Courts Provincial Criminal Fine Sentences, Courts Provincial Criminal Other Sentences, Courts Provincial Criminal Total Sentences, Total Sentences as a percentage of Guilty Findings.

- **The Corrections Data Fields**

    1. Monthly Admissions to BC Corrections Data: Alternative Measures, Recogni-
       zance/Peace Bond, Bail, Remand, Custodial Sentences, Probation, Conditional
       Sentences.
    2. Average Daily Counts Per Month: Alternative Measures, Recognizance/Peace
       Bond, Bail, Remand, Custodial Sentences, Probation, Conditional Sentences.

5.2 Verification and Validation of the Model

Verification is the process of ensuring that the model behaves as intended. Throughout
the development phase, we constantly checked the elements of the model with the
system, to ensure it behaves as intended. One of the verification check performed
on this model was a steady-state analysis. In the absence of changing influences, we
would expect that over time the average number of RCCs, people, and cases in each
part of the system would approach to a constant value. This equilibrium arises because
competing influences in a stable system eventually reach a balance. Verifying that
the model exhibits this behaviour demonstrates that the key competing influences
are incorporated in the model. Steady state analysis requires the development and
analysis of differential equations for each stock. It was evident that the stocks in the
model for which this analysis is conducted reach a steady state rapidly.

After verifying step, we validated the model. Validation ensures that no significant difference exists between the model and the real system. This is an important
step to show that the model reflects reality and can be used to accurately address
"what-if" questions, and to forecast the future behaviour of the system over time.
We used historical system data to develop the mathematical functions and formulae
to represent the behaviour of the stocks incorporated in the system dynamics model.
The results are then tested against real data to determine the level of the *fit* between
the simulation results and the reality from the data. A good *fit* make us confident in the
quality of the future predictions derived from the changes to the model parameters.

6 Simulated Scenarios

Here, we illustrate how the model can assist policy makers and mangers to assess
potential inteventions by simulating them before executing them. In this section,
we present two different scenarios and compare each to the status quo. Scenario 1
involves changes to UCR clearance practices; Scenario 2 involves changes to the
volume of UCR offences reported to the police.

We were required to keep all data confidential. Accordingly, the vertical scales
on all graphs have been eliminated.

6.1 Scenario 1: The Development of a Large Task Force

An integrated gang homicide task force is created in the metro Vancouver area of British Columbia in June of Year 1 to respond to a recent surge of gang-related homicides. As a result, police agencies in the Lower Mainland are under pressure to close as many open cases as possible and to free up resources for assignment to the task force. The result is that clearance rates double (non-cumulatively) in the months of June, July, August and September of Year 1. The simulation of this scenario illustrates the predicted downstream effect on the number of accused offenders in remand when the number of offenses cleared by charge doubles in each of the four months in question.

Figure 8 shows the "intervention" being simulated, that is, the flow of offences cleared by charge under the status quo and when such activity doubles during the four month period.

Figure 9 shows one downstream effect of the intervention, the impact on the number of accused offenders in remand. Figure 9 illustrates how a short-term change in an upstream activity (offences cleared by charge) has a prolonged effect on a downstream variable (accused offenders in remand). The model projects that the doubling of clearances over four months results in a significant increase in offenders in remand over an entire year. Over this year, on average, there are about 9 % more offenders in remand as compared to the status quo. The strongest effect is in September of Year 1 when about 20 % more offenders are in remand.

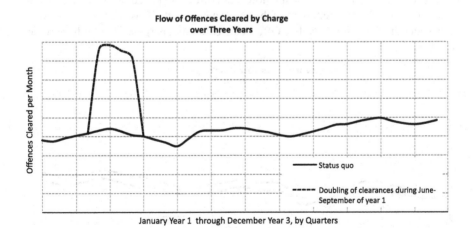

Fig. 8 A four month increase in UCR cleared by charge events

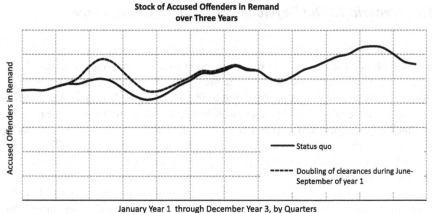

Fig. 9 The impact on remand of a four month increase in UCR cleared by charge events

6.2 Scenario 2: A Broken-Windows Policy

In January of Year 1, police agencies in across British Columbia embark on a strategy that incorporates the "broken windows" thesis, targeting drug, mischief, and other property damage offences accordingly. Officials estimate that the strategy will lead to a 40 % increase in UCR reported offences of these types over the subsequent three years. The simulation of this scenario illustrates the predicted downstream effect on (all) offenses cleared by charge and accused offenders on bail.

Figure 10 shows the effect on the flow of all offences cleared by charge under the status quo and when the reporting of the above mentioned offenses increases by 40 % every month (non-cumulatively). Offences cleared by charge each month increase by about 12 %.

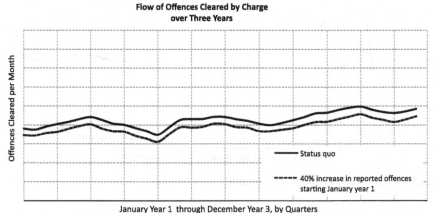

Fig. 10 The effect a 40 % increase in drug, mischief and other property damage offences on the number of UCR cleared by charge events

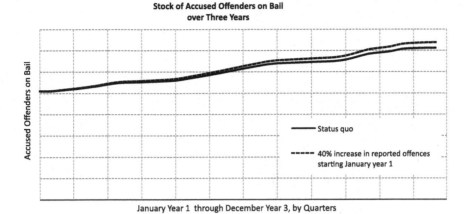

Fig. 11 The impact of a 40 % increase in drug, mischief and property damage offences on the number of accused on bail

Figure 11 shows the effect of 40 % more drug, mischief and other property damage offences being reported on the number of accused offenders on bail. The impact on the stock of accused offenders on bail is cumulative. 40 % more reports of the above mentioned offences leads to about 12 % more offenses cleared by charge which leads to a cumulative 1.8 % per month increase in offenders on bail.

7 Conclusions and Future Works

This chapter is restricted to the development of a high-level model of a CJS containing the main sub-systems police, Crown, court and corrections. Clearly it is very important to incorporate *queues* in the model to more accurately represent "wait-time" behaviour. This provides a more accurate method for estimating wait-time behaviour and delay in a CJS. In the model described in this paper resourcing is not incorporated so in future research resourcing model can be developed for the different sub-systems.

The current model is *linear*, meaning it possesses no feedback. Feedback is a critical characteristic of dynamical systems, such as CJSs. Feedback occurs when certain stocks have limits that, when approached, necessitate a system response or adaptation. To accomplish this requires the identification of the "stocks" in the model where these limits exist. Subsequently, the stocks that are the most crucial to system functioning can be determined. The next step in this process necessitates an understanding of the types of adaptations that can and do take place when a critical stock reaches capacity. This means that in the process a *qualitative* description of which other "stocks" change their behaviour in response to a capacity limit of a critical stock is pivotal. This qualitative description, which should be gleaned from the expertise

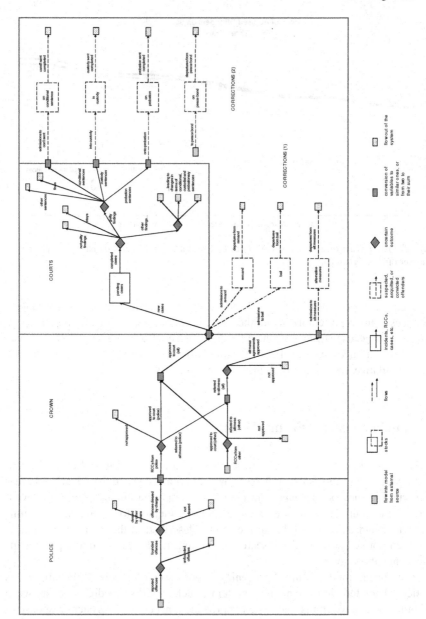

Fig. 12 Flowchart of the B.C. criminal justice system

of system managers, can then be transformed into a mathematical representation of feedback behaviour. An example might be, what happens at the various stages in a CJS when *Pending Cases* reaches a limit?

The model in this paper deals with overall UCR reported offences. As a result, there is no prioritization of UCR reported offences in this model. It is important that a prioritization scheme be implemented to reflect the dynamics of the input data. This prioritization scheme could be set at a general level (i.e., summary conviction/indictable or major crime groupings — violent, property, drug and other) or a highly specific level of detail (homicide offences, assault offences, robbery, other violent offences, theft offences, break and enter offences, mischief offences, drug possession offences, drug trafficking offences, administrative offences, etc.). This type of prioritization scheme allows the model to make more detailed predictions and it enhances the detail of the "what-if" questions that may be tested.

As another future work we can assign different attributes to the entities that impact their flow through a CJS. For example, age of the offender, gender of the offender, offence location and risk classifications (i.e., low, medium, high) of offenders on probation impact the outcomes and paths in the system. These are important influences in a detailed model of the criminal justice system in British Columbia, and have a different effect on resource utilization.

Acknowledgments This work was supported by the Complex Systems Modeling Group (CSMG) at the Interdisciplinary Research in the Mathematical and Computational Sciences (IRMACS) Centre - Simon Fraser University, and by the working group and steering committee of the Ministry of Public Safety and Solicitor General and the Ministry of Attorney General (PSSG/AG) in British Columbia, Canada. We would like to acknowledge the input from the members of the Complex Systems Modelling Group. We also are grateful for technical support from the IRMACS Centre, Simon Fraser University.

References

1. Alimadad, A., Borwein, P., Brantingham, P., Brantingham, P.J., Dabbaghian-Abdoly, V., Ferguson, R., Fowler, E., Ghaseminejad, A.H., Giles, C., Li, J., Pollard, N., Rutherford, A., van der Waall, A.: Using varieties of simulation modelling for criminal justice system analysis. In: Eck, J., Lui, L. (eds.) Artificial Crime Analysis Systems: Using Computer Simulations and Geographic Information Systems, pp. 372–412. Hershey: Idea Group Inc., USA (2008)
2. Bailey, J., Hann, R.: Systems Analysis and the Corrections System in Ontario. Centre of Criminology, University of Toronto, Toronto (1972)
3. Belkin, J., Blumstein, A., Glass, W.: JUSSIM, an interactive computer program for analysis of criminal justice systems, Working Paper. Carnegie-Mellon University, Urban Systems Institute, School of Urban and Public Affairs, (1971)
4. Blumstein, A.: Crime modeling. Oper. Res. **50**(1), 16–24 (2002)
5. Blumstein, A., Cassidy, G., Hopkinson, G.: Systems analysis and the Canadian justice system, CANJUS Project Report No. 14. Ottawa, Ministry of the Solicitor General (1975)
6. Blumstein, A., Cassidy, G., Townsend, J.: Implementation of a systems approach to the Canadian criminal justice system, Statistics Division Report No. 3. Ottawa, Ministry of the Solicitor General (1974)

7. Blumstein, A., Larson, R.: Analysis of a total criminal justice system. In: Drake, A., Keeney, R., Morse, P. (eds.) Analysis of Public Systems. Massachusetts Institute of Technology, Boston (1972)
8. Brantingham, P.: Dynamic modeling of the felony court system. Unpublished doctoral dissertation,Florida State University (1977)
9. Brantingham, P., Brantingham, P.: Computer simulation as a tool for environmental criminologists. Secur. J. **17**(1), 21–30 (2004)
10. Comfort, J., Shapiro, S., Volcansek-Clark, M.: Development of a Simulation Model of Criminal Caseflow Activities, Decision and Policy Variables. Florida International University, Miami (1982)
11. Dabbaghian, V., Fowler, E., Giles, C., Richardson, N., Rutherford, A.R., van der Waall, A.: Simulation modelling of the criminal justice system: a high-level model for british columbia. In: Report for the British Columbia Ministry of Public Safety and the Solicitor General and the Ministry of the Attorney General, Canada (2007)
12. Hann, R., Bailey, L., Ross, M.: Decision Making in the Canadian Criminal Court System: A Systems Analysis, Volume I and II. Centre of Criminology, University of Toronto, Toronto (1973)
13. Hann, R., Salzman, L.: CANCOURT - I: A Computerized System Simulation Model to Support Planning in Court Systems. Decision Dynamics Corporation, Toronto (1976)
14. Larson, R., Cahn, M., Shell, M.: Improving the New York City arrest-to-arraignment system. Interfaces **23**(2), 76–96 (1993)
15. McAllister, W., Atchinson, J., Jacobs, N.: A simulation model of pretrial felony case processing: A queuing system analysis. J. Quant. Criminol. **7**(3), 291–314 (1991)
16. McGinnis, J.: Predicting police force attrition, Promotion and demographic change: A computer simulation model. Can. Police Coll. J. **13**(2), 87–127 (1989)
17. Stollmack, S.: Predicting inmate populations from arrest, Court disposition and recidivism rates. J. Res. Crime Delinquency **10**(2), 141–162 (1973)

Chapter 7
Celerity in the Courts: The Application of Fuzzy Logic to Model Case Complexity of Criminal Justice Systems

Andrew A. Reid and Richard Frank

Abstract There are many complex phenomena in the criminal justice system that are difficult to understand because they contain features or concepts that are fuzzy in nature; in other words, it is difficult to assign a crisp value or label to them. Fuzzy Logic is a mathematical concept developed to deal with these very problems. Fuzzy Logic techniques are capable of dealing with approximate facts and partial truths, and not just precise values to model complex issues and processes. While Fuzzy Logic has been used in a number of criminology and criminal justice research efforts, it has not been applied to issues in the criminal court system. Case management is critical to ensure court systems run efficiently and understanding case complexity is an important part of that task. In this chapter we propose Fuzzy Logic as a technique that could be used to model the complexity of cases using general characteristics that are known before cases enter the court system. Using the adult criminal court system in British Columbia as an example, we propose a model that could predict case complexity immediately following the laying of charges by Crown prosecutors. By understanding case complexity a priori, courts may be able to enhance early case consideration procedures such as screening and scheduling to create a more effective and efficient justice system.

1 Introduction

Criminal justice systems involve many complex relationships. Police, Crown, courts, and corrections are some of the key components of these systems that must work together to deliver justice when breaches of criminal law occur. While the

A. A. Reid (✉) · R. Frank
Institute for Canadian Urban Research Studies, School of Criminology, Simon Fraser University, Burnaby, Canada
e-mail: aar@sfu.ca

R. Frank
e-mail: rfrank@sfu.ca

V. Dabbaghian and V. K. Mago (eds.), *Theories and Simulations of Complex Social Systems*, Intelligent Systems Reference Library 52, DOI: 10.1007/978-3-642-39149-1_7, © Springer-Verlag Berlin Heidelberg 2014

inter-relationships between these groups pose considerable challenges to maintaining a functioning system, each too, is taxed with its own set of challenges. The criminal court system is perhaps faced with one of the greatest challenges. Criminal courts must schedule appearances for individuals accused of offences in conjunction with other key players such as judges, Crown and Defence counsels, sheriffs, witnesses, and victims while following strict procedural rules. This is particularly challenging in urban centres that deal with large caseloads. Managing large case volume while dealing with factors such as reduced funding and increasing costs, make it difficult for criminal courts to maintain expeditious case processing and as a result, delays may occur.

One of the factors that is closely linked to delay in the courts is case complexity. Understanding case complexity is critical because different scheduling and case-flow management strategies are needed to effectively and efficiently process all types of cases through the system. One major barrier to understanding case complexity is that there is no clear definition for what constitutes a complex case, nor is there a series of clearly defined complexity measures upon which cases may be classified. Using the justice system of the province of British Columbia, Canada, as an example, this chapter introduces the longstanding problem of delay in criminal court systems and proposes a mathematical modeling strategy that could be used to develop an early case consideration tool. Specifically, this tool could be used to identify the complexity of cases based on characteristics that are known before they enter the court system.

2 Case Processing Patterns in Criminal Courts

2.1 Causes of Delay

While there are processes in criminal courts that may delay the progression of cases, many are either necessary or unavoidable [34]. Nevertheless, these legitimate delays contribute to longer case processing times and should not be ignored. The length of criminal trials is one legitimate factor that may impact case processing times in the courts. Although many criminal justice professionals believe that a lengthy trial process may be the natural result of a complex case, there is increasing evidence that the length of trials, in general, is increasing. In a recent speech, the Right Honourable Beverley McLachlin noted that national trends of trial lengths in Canada have been on the rise. In the recent past it would not be uncommon to have murder trials last 5–7 days and now they more frequently last 5–7 months [21].

Unnecessary delays, however, are also a concern because they may further contribute to a number of unfavourable consequences. These types of delays may stem from a variety sources. Adjournments are one example. An adjournment is the suspension of legal proceedings to another time or place. Unnecessary adjournments may occur when a hearing has been scheduled but one or more of the parties are not prepared to participate. These ineffective hearings often lead to a motion for adjournment so the hearing can take place at a later time or date. This type of defer-ral has long been recognized as a factor contributing to delays in criminal justice systems and a recent study suggests that the problem continues to plague court systems [12].

A lack of resources may also contribute to unnecessary delays in the courts. Over the past several months there has been growing concern throughout BC that a shortage of Provincial Court staff may result in significant delays in criminal courts [11]. In a meeting between BC Supreme Court Chief Justice Robert Bauman, Attorney General Barry Penner, BC Chief Justice Lance Finch, and Provincial Court Chief Judge Thomas (Tom) James Crabtree, the group expressed grave concern about trials being delayed due to budget cuts [11]. Budget cuts may not only lead to fewer cases being scheduled for a given period but also to unanticipated adjournments if staff are unavailable to hold a hearing on the date for which the proceeding is scheduled.

Case complexity is another factor that has been found to contribute to delays in criminal court processes. While there is no concrete definition for complexity, there are some common characteristics that contribute to a case being defined as more or less complex. First, the type of charge has been noted as a key factor. Generally, the more serious the charge, the more complex the case is likely to be. For example, cases that include charges of aggravated and sexual assault, break and enter, and dangerous driving will be scheduled for longer trial time [31]. The number of participants in the court matter has also been linked to case complexity. A case involving multiple accused, for example, is likely to add to the complexity because the case may require decisions of joinder[1] or severance of charges[2] [6]. Similarly, cases involving participants with extraordinary circumstances such as a mentally ill defendant or defendant requiring the consultation of separate legislation (E.g., a child or youth is associated with an adult criminal case) are likely to be more complex.

2.2 Consequences of Unnecessary Delay in the Courts

Unnecessary delays in criminal court systems may have serious implications that affect many strata of society. From a social cost perspective, delays in case processing may lead to inappropriate or undesirable conclusions in legal proceedings. In Canada, for example, any person charged with an offence has the right to be tried within a reasonable time [5]. If this right is denied, a person charged with a criminal offence may make a Charter challenge and have their charge stayed[3] or withdrawn.[4] While

---

[1] Joinder is a legal term that refers to the process of merging two or more legal matters to be heard in a single trial or hearing.

[2] Severance of charges refers to a process where two or more related offenses are charged in a single indictment or information [27].

[3] A stay or stay of proceedings, refers to the suspension of a charge, either temporarily or indefinitely. In Canada, charges that have been stayed may be resumed in court within one year.

[4] While there are many instances where stays or the withdrawing of charges are used for legitimate reasons that pertain to matters of a case, issuing a stay of proceedings or withdrawing a charge due

the terminology used to declare this right is rather vague, and the Supreme Court of Canada has not defined a strict limit for time to conclusion, guidelines have been established in case law that affirms most cases should be completed within an 8–10 month period from when the charges were laid [6]. The Askov [26] landmark case that established the interpretation of Section 11(b) of the Charter provides a clear example of how delays in judicial processing can result in undesirable conclusions to cases. In the year following the Askov [26] decision, 50,000 charges in Ontario were either stayed or withdrawn pursuant to Section 11(b) of the Charter [16].

Unnecessary delays may also lead to unreasonable infringements on the rights and freedoms of persons charged with criminal offences. With respect to the former, persons awaiting trial may be subject to lengthy periods of incarceration if they are remanded[5] in custody and delays prevent the case from moving forward. This is particularly concerning in cases where an accused is remanded and spends longer in pre-trial custody than would be an appropriate sentence issued after a conviction on the original offence. Instances like this may occur if an accused is originally granted bail and subsequently violates one of the terms of their release. This could result in the accused being held in custody for breaching bail conditions when the original offence may not have warranted the length of custodial sentence already served in pre-trial detention.

Another consequence of delay in court is that it may impact the quality of legal arguments in cases. For example, if trials are delayed for extended periods of time, evidence may be lost or damaged. In Canada's adversarial legal framework, this could seriously impact the strength of cases delivered by Crown or Defence counsels. Further, trial delays may cause witnesses to become less reliable when requested to testify to events. Similar to the reliability of evidence, the credibility of witnesses may be deflated and impact the quality of cases put forward by legal counsels. Both of these consequences could potentially result in injustices either to the accused or the State.

From an economic perspective, delays may impact case processing and consequently lead to more resource demands on the justice system. For example, delays may result in longer case processing times and extended pre-trial custody holdings due to increases in bail hearings, constitutional challenges, and other court appearances that cause interruptions to the expeditious processing of a case. With these types of case processing impediments, resources of the police, crown, courts, and corrections may be taxed. As a consequence, federal and provincial governments may have to bear greater economic burdens to maintain a functioning justice system [21].

(Footnote 4 continued)
to an unreasonable delay is considered an undesirable conclusion because the cause of the dismissal is beyond the parameters of the court matter; the facts of the case nor the interpretation of criminal law pertaining to the charges are at question in these particular instances.

[5] Remand is a form of pre-trial detention that may be used when incarceration is deemed necessary in order to ensure the accuseds attendance at trial, to protect the public, or to prevent the administration of justice from falling into disrepute [30].

2.3 Recent Trends of Delay in Court

There have been a number of important efforts that have attempted to reduce or eliminate unnecessary delays in criminal justice systems. During the 1970s through the 1990s there was major reform in North American court systems that introduced initiatives such as court delay evaluations, court delay reduction programs, court management programs, and research tools for examining justice system processes (see for example [3, 8, 9, 23, 28]). In Canada, specifically, the R. v. Askov case prompted a series of initiatives to improve case processing and reduce the length of time between initiation and disposition. For example, s. 625.1 of the Criminal Code permitted pre-trial conferences and s. 482.1 allowed courts to make rules that would aid in effective and efficient case management [16].

Despite these initiatives, delay in criminal courts is still a major problem. In recent years, criminal court systems across Canada have demonstrated marked declines in expeditious case processing. In fact, recent analyses of national case processing times in adult criminal courts reveal a general upward trend where in 2000, the median elapsed time for case completion was 101 days; by 2009 this number had risen to 124 [32]. BC's criminal court system is an exemplar of this issue. The number of cases entering the criminal courts have been decreasing for some time, yet the number of appearances per case, the length of time to disposition, and the number of pending cases in excess of 240 days all reveal an increasing trend [1]. These delays, coupled with case complexity, have been identified by both the Right Honourable Beverley McLachlin [21], Chief Justice of Canada and former Chief Justice Antonio Lamer [15], as being among the greatest challenges facing the justice system.

Clearly, there is still room for much improvement in reducing delay and expediting cases through the criminal court system. Encouragingly, there is a general consensus on areas that should be targeted for contributing to effective case flow management. Among others, these include effective early case screening and vetting by prosecutors, case differentiation with separate tracks used for the processing of cases depending upon their complexity, case scheduling practices that facilitate expeditious disposition of simple cases, and rapid identification of cases likely to require more counsel time and judicial attention so that good use can be made of limited courtroom capacity and counsel preparation time [13].

Underlying all of these areas is the opportunity for an early case consideration tool that can predict the complexity of cases before the case is initiated in the court. The prediction of case complexity, however, is a challenging task. There is no clear cut definition of complexity or available measures by which it can readily be calculated. As a result, complexity is generally thought of on a continuum between the most simple types of cases and the most complex. Fuzzy logic provides an opportunity for these imprecise concepts to be expressed in mathematical terms.

3 Fuzzy Logic

First introduced by Lotfi Zadeh in 1965 [19] in the proposal of fuzzy set theory, Fuzzy Logic is a form of logic that allows for reasoning that is approximate rather than crisp or exact. For most cases in mathematics, crisp values are required. The sum of 2 and 2 will always be 4; there can be no fuzziness about that. However, humans are not always so logical or precise. When it comes to modeling people and human behaviour, this uncertainty can well be captured by Fuzzy Logic. This is particularly true with respect to criminal court systems where many factors may influence case processing; precedence must be taken into account and foresight must be used since the process of hearing a case can set precedence for future cases yet to come. These factors introduce a lot of uncertainty into decisions that are made. This makes linear modeling techniques less appropriate and the application of Fuzzy Logic ideal.

Contrary to what the terms seems to mean (that the logic is vague or uncertain), fuzzy logic actually follows a precise set of rules and fuzzy simply refers to the fact that this type of logic can deal with vague or uncertain input and output variables. Traditional logic is binary, either true or false. However, in the real world, this is usually not the case, with many decisions being vague, or fuzzy. Terms like tall and short can be used to describe the height of people with, for example, people over 2 m's considered tall and people shorter than 1.5 m considered short. While most people would agree that 1.9 m is tall, and 1.6 m considered short, as the height under consideration increases from 1.6 to 1.75 m, there will be a point when the decision to call it short or tall will not be unanimous anymore; some people will call it short and some will call it tall. That is, the label assigned to it starts to become fuzzy.

3.1 Fuzzy Logic Techniques

Since fuzzy logic was developed, it has been used in various systems, from vibration damping, balancing of an inverted pendulum, to machine scheduling [24]. Although the logic itself is able to deal with vague or fuzzy terms, the underlying model is very precise. There are five stages to the Fuzzy Inference System (FIS) that makes up the fuzzy model. This is shown in Fig. 1, and detailed below.

Stage 1—Input Variables. The first stage of the FIS is the collection of the data for the input variables. For the application proposed here, this is discussed in greater detail in a simple example in the section titled Applying the Model to Court Case Complexity. The input variables at this point come directly from the data source, and hence are crisp values. They must be put through the fuzzifier (stage 2) before they can be used by the fuzzy system.

Stage 2—Fuzzifier. The fuzzifier turns the crisp values into linguistic variables, such as *Low*, *Medium* and *High*, as required by the Inference System. The fuzzifier for a single variable is composed of multiple membership functions that take a crisp

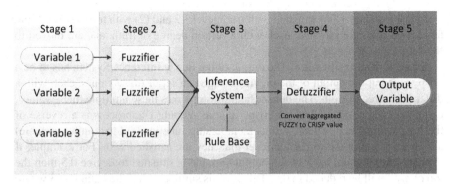

Fig. 1 Fuzzy inference system

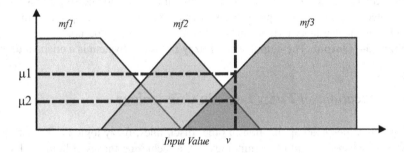

Fig. 2 Fuzzifier

value, and possibly assign multiple linguistic variables to it with different truth values (μ) depending on which membership function is in effect. For example, a fuzzifier with three membership functions is shown in Fig. 2. The input variable that has value v potentially activates membership function $mf2$ and $mf3$. For this particular value of v, $mf2$ has a truth value $\mu2$ while $mf3$ has truth value $\mu1$.

Stage 3—The Inference System. The Inference System takes the linguistic variables and applies a pre-determined set of rules, designed based on domain knowledge or derived from data, which connect the input linguistic variables to output linguistic variables. The rules are defined in the format of IF/THEN implications where the antecedents are composed of input linguistic variables connected via AND, OR and NOT Boolean operators, and the consequent refers to one of the values of the linguistic output variable. The Inference System takes the values of the input linguistic variables and, based on the rule-set, assigns a value to the output linguistic variable. In most circumstances multiple rule-sets will be satisfied, each with some level of truth. For example, assume $InVariable2 = mf3$, then the antecedents of the following rules could be satisfied:

(1) IF $InVariable1 = mf2$ THEN $OutValue = mf5$
(2) IF $InVariable1 = mf2$ AND $InVariable2 = mf3$ THEN
$OutValue = mf4$

In this instance, (1) is satisfied with truth level $\mu2$ and (2) with truth level $\mu1$ and hence two rules are satisfied, each with a certain degree of truth, and are moved to the defuzzification step.

Stage 4—Defuzzification. The output of the Fuzzy Inference System has to be a crisp value but the output of the previous stage is a set of fuzzy values, each with a different degree of truth. The defuzzification step takes these multiple fuzzy values, combines them, and produces a crisp value as output. The process is a reverse of the fuzzification step, where various membership functions (of the output variable) are *chopped off* at the top at the level that they are satisfied with. For example, if membership function $mf3$ of the output variable is satisfied to degree 0.5 then the top of $mf3$ will be reduced to 0.5. The result is similar to that shown in Fig. 3 where $mf1$ is satisfied to degree $\mu1$ and hence chopped to that level. The same process applies for $mf2$ and $mf3$. Then, the total area under all membership functions is aggregated and the crisp output value v is calculated based on a specified method usually center-of-gravity (which splits the area into two equal parts).

Stage 5—Output. The output of the Fuzzy Inference System is a crisp value v.

3.2 Applications of Fuzzy Logic in Criminology

Despite its widespread application in other disciplines, fuzzy logic has been rather underutilized in the field of criminology. As a technique that is able to deal with vague and non-crisp phenomena, however, fuzzy logic appears to have many potential applications in research on crime and delinquency. In policing, for example, Verma [33] noted many areas where fuzzy logic could be applied:

> Police officers commonly receive descriptions of suspects that are fuzzy in nature. Offenders are described as "tall," "dark," "young" or even "rude," terms that are imprecise and admit a range of possibilities. In fact, policing itself involves many issues that are fuzzy and difficult to measure exactly. For example, officers' services are often evaluated as being "good" or "average" while gang-activity -related areas are described as "dangerous" or simply "rowdy." Police managers constantly strive for "appropriate" resources and "better" training facilities, while the public expects the police to be "honest" and keep the streets "safe." All these characteristics are essentially fuzzy and therefore difficult to assess through common statistical techniques [33].

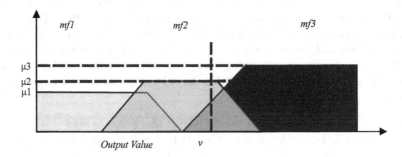

Fig. 3 Defuzzifier

Extending this idea, Verma [33] introduced common mathematical techniques used in fuzzy logic applications and provided a hypothetical example of fuzzy logic in the context of policing. In offender profiling, information about potential suspects is gathered from a variety of sources and much of that information is vague (particularly if it is collected from eye-witnesses). More accurate profiles of offenders may be able to be gleaned, Verma [33] argued, when fuzzy logic techniques are employed.

Grubesic [10] introduced fuzzy logic techniques to tackle another problem in criminology-the detection of spatial crime patterns. In the clustering of crime observations into hot-spot groupings, there are often instances where observations do not clearly fall into one group or another. As a result, it may be difficult to justify the rigid statistical rules that determine the clusters. Through the use of a fuzzy clustering technique, however, Grubesic [10] was able to demonstrate how the approach was better suited than other hard-clustering techniques, to deal with outliers and cases that were not clearly able to be assigned to a cluster.

Fuzzy techniques have also been employed to develop effective law enforcement strategies and optimal deployment of crime prevention resources. In Li et al. [17], for example, the authors note that previous research on crime policy has primarily employed data mining techniques. Although these have proven to be useful, they are limited with respect to their ability to incorporate linguistic forms of data. As a result, they used a fuzzy self-organizing map strategy to investigate temporal patterns in various crime-type categories. Through the use of this technique, Li et al. [17] were able to discover four distinct temporal crime pattern trends. The results, they argued, could be used to more effectively and efficiently deploy police resources for tackling specific crime problems.

Criminologically speaking, not all locations are created equal, with some locations experiencing high levels of crime while others experince very little. However, the location of crimes are partially driven by the travel patterns of offenders, who, when travelling towards certain locations (termed criminal attractors), become aware of criminal opportunities there. This is described in Crime Pattern Theory [4]. In Mago et al. [20], the authors applied Fuzzy Logic to this theory in order to model the movement of offenders towards known criminal attractors (in this case shopping centers) within several cities. The authors took the known home location of each offender, and based on the offender's crime locations, used Fuzzy Logic to determine which shopping center the offender was moving towards, when they stopped along the way to commit their crimes. The model successfully provided results that were very comparable to real life expectations.

Cyber-crime is an ever-increasing focus of criminological research because until recently, the online environment had gone virtually unsupervised. With the ability to communicate, share information and conduct transactions in an online environment, however, the opportunities for crime are immense. As a result, investigative tools that may be used online are of great importance for those interested in preventing and reducing cyber-crime. One barrier to effective online investigations is that evidence may not be able to be properly retrieved. In an attempt to address this problem, Liao et al. [18] developed an approach based on fuzzy logic and expert systems that can create digital evidence of cyber-crimes automatically. In their analyses, Liao et al. [18] found their fuzzy logic-based method of cyber-crime evidence collection, organization, and presentation to outperform comparable methods.

In a different cyber application, Botha and Solms [2] proposed a fuzzy logic-based method that could be used to minimize and control intrusion attempts to an organization's computer system. The program they developed used fuzzy logic to determine if an intrusion attempt had taken place and if so, to what extent the intruder was successful in gaining access to the computer system. The program was developed so that security officers could assess the level of security of their computer systems and obtain information about potential threats. A similar tool was also proposed by Mohajerani et al. [22]. Taking this preventative approach to cyber-crime has demonstrated the versatility of fuzzy logic techniques to tackle crime issues that may occur in an online environment. Applications in other sectors of criminal justice, however, show equal promise. Verma [33] emphasizes that many issues in the field of criminal justice are fuzzy in nature and fuzzy logic techniques show potential for applications in law and legal research.

3.3 Applying the Model to Court Case Complexity

In order to identify the complexity of cases in the court system, a model centered around Fuzzy Inference Systems (FIS), called the *Complexity Estimator FIS* (CEFIS), shown in Fig. 5, could be developed to calculate the complexity of court cases. Complexity of a court case may depend on many factors, and all of these factors could be used as input into CEFIS. For illustration purposes, assume that complexity is calculated on only three factors: (1) number of charges contained in the case, (2) number of people in the case, and (3) the severity of the charges. The output of CEFIS would be the complexity of the court case; and therefore, the higher the Complexity Measure, the more complex the case should be expected to be. For example, assume a court case involving four accused persons is initiated in court and each person has a single charge against them (thus four charges in total) for shoplifting (severity level 1). The complexity expected for this specific court case would be expected to be somewhat high due to the number of people involved, and the number of charges contained within the case. Here, severity would not play a large role since it is a relatively non-severe crime.

Using these assumptions, the FIS would contain the 5 stages, shown in Fig. 4:

Stage 1—Input Variables. The inputs used in CEFIS are composed of three variables:

1. The first parameter, *Number of People*, denotes the number of accused people associated to the case, and would be represented as a positive integer.
2. The second parameter, *Number of Charges*, denotes the number of charges brought against all of the people associated to the case. This parameter is also represented as a positive integer.

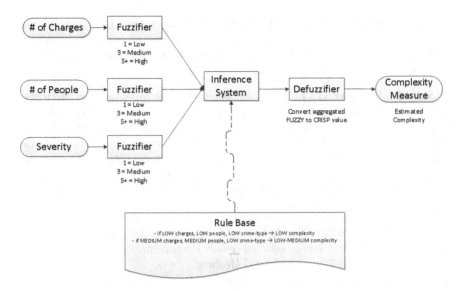

Fig. 4 Complexity estimator fuzzy inference system

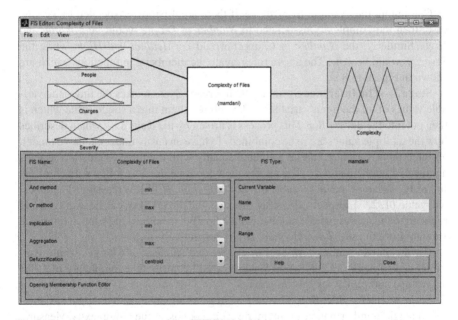

Fig. 5 Possible implementation of the CEFIS in Matlab

3. The final input parameter, *Severity*, is more difficult to define since the crime types do not categorize into positive integer values. For this parameter, each crime type could be translated into an integer representing the severity of that specific crime

(for example, via the use of a Crime Severity Index[6]). For cases with multiple charges of varying degrees of severity, the average severity of the charges could be calculated and used as the input into CEFIS. In the example from above, the input parameters would be 4, 4 and 1, respectively.

Stage 2—Fuzzifier. The fuzzifier step takes each of the input variables and translates them to a categorical value through the use of a membership function. Due to the lack of research on fuzzy logic applications in criminal court systems, the exact membership function for each of the parameters would have to be based on qualitative analysis of court cases or in speaking to those intimately familiar with the process, such as judges or judicial case managers. For the demonstrative purposes of the model presented in this chapter, the linguistic variables in the three membership functions are formulated as follows:

- if the input variable has value 1, then its contribution to complexity is *Low*
- if it has input value 2, it is considered equally *Low* and *Medium*
- if it has input value 3, it is considered *Medium*
- if it has input value 4, it is considered equally *Medium* and *High*
- if it has input value 5, it is considered *High*.

Continuing the example from above, if there are 4 people associated to a court-case, then the complexity associated to *Number of People* would be *Medium* and *High*. Similarly, the *Number of Charges* would be *Medium* and *High*, while the *Severity* would be *Low*. The above rules create the membership functions which are shown in Figs. 6 to 8.

Stage 3—The Inference System. The Inference System takes the linguistic variable values of all the input variables and combines them into a single value which is then put into the defuzzifier. This process is driven by the Rule Base which describes how the multiple values are to be combined. To have an easily understandable model, all three input variables can be weighted equally with the three linguistic variables aggregated mathematically. The aggregation could be done as follows. Each of the linguistic variables, having a value of {*Low, Medium, High*}, is reassigned a numerical value {1, 2, 3} respectively. After this step, the three values can be averaged and rounded to the nearest whole number, which finally is assigned back into a linguistic variable from the set {*Low, Medium, High*} respectively.

For the above example, *Number of People* and *Number of Charges* would contribute a complexity of 2.5, while *Severity* would contribute a complexity of 1, for an average complexity of 2, or *Medium*.

Stage 4—Defuzzification. This step aggregates all values of the output variables into a single number which in the model corresponds to the Complexity Measure. All membership functions which are satisfied during the previous stage contribute towards this value. The Complexity Measure can be calculated by taking the membership function that corresponds to the output value of the Inference System, and

---

[6] In Canada, a Crime Severity Index was developed by Statistics Canada to offer an alternative measure of police reported crime [29].

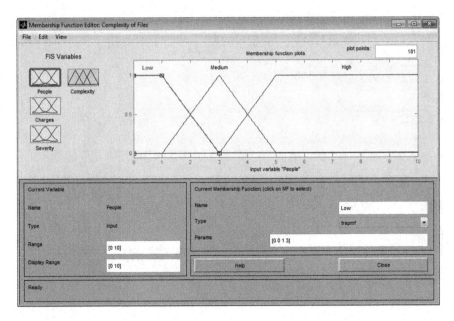

Fig. 6 Input variable *number of people*

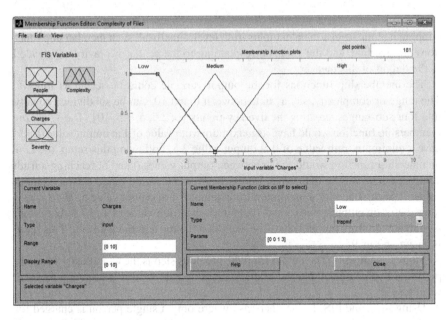

Fig. 7 Input variable *number of charges*

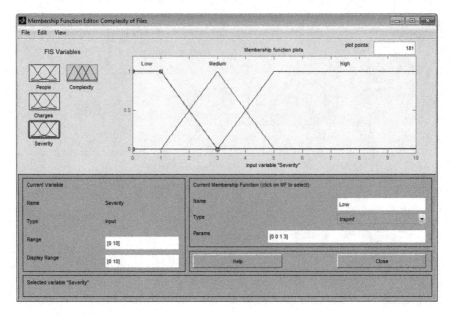

Fig. 8 Input variable *severity*

clipping it at the level equivalent to its truth level. Finally all such membership functions are overlaid, and the value corresponding to the center of gravity is provided as the output of this measure.

The membership functions for the output variable could be set up as follows. The range or complexity, say a value between 0 and 10, can be subdivided equally into four sub-ranges, creating the five key-points {0, 2.5, 5, 7.5, 10}. The *Very Low* membership function would have its maximum truth value of 1 at output value 0, and have a minimum truth value of 0 at output value 2.5. Following this setup, the *Low* membership function would range between output values 0 and 5, reaching a truth value of 1 half-way in that range. Similarly, *Medium* would range between output values 2.5 and 7.5, with a maximum truth value of 1 at output value 5. *High* would range between output values 5 and 10, and reach a truth value of 1 at 7.5. *Very High* would then go from output value 7.5–10, with it having a truth value of 0. This setup is shown in Fig. 9.

Stage 5—Output. The output from the previous step is the Complexity Measure produced by this model. For the example presented above, this complexity measure would be 5.

Using the same FIS, for another case where only a single person is charged for a single crime of severity one, it would be expected that the complexity would be much smaller due to there being only a single person involved. Indeed, the above FIS does produce a complexity of 0.8 (as opposed to 5 for the previous example). Due to the structure of the FIS however, the values yielded by the FIS cannot be compared

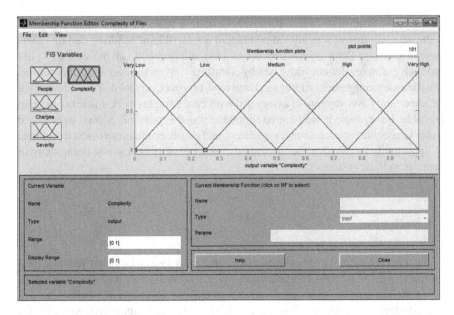

Fig. 9 Output variable *complexity*

linearly, thus a case with estimated complexity of 5 is not 6 times as complex as one with complexity of 0.8.

3.4 Utility of a Fuzzy Logic Case Complexity Tool

This example demonstrates how knowledge of some basic characteristics of cases could be used to predict the complexity of cases before they enter the criminal court system. More refined and accurate prediction models show the potential to invoke early case intervention strategies that could aid in better case management. Complexity is one of many important elements that is known to impact case processing in criminal court systems. When the complexity of cases is not understood or managed effectively, expeditious case processing may be impacted, resulting in delays. Delay may lead to a number of unfavourable outcomes including charges being stayed or withdrawn, an accused being held in custody for an unreasonable time, the deflation of witness testimony, and greater economic burdens for government bodies that fund judicial systems. Better understanding case complexity, however, may allow case flow management strategies to improve case processing. Being able to classify cases by their respective complexity a priori, for example, may allow for effective early case screening and vetting, case differentiation for creating case flow tracts, improvements to case scheduling practices, and rapid identification of cases likely to require more time and resources.

The identification of (administratively) different types of cases has already led to the development of specialized courts that are capable of handling certain issues more effectively. Early disposition courts [14] and problem solving courts [7] are two such examples. Identifying cases by complexity opens the door to further refined judicial case management strategies. Until now, however, no tool that can account for the uncertainty and vagueness associated with case complexity definitions has been available. Fuzzy logic is well-suited for addressing this problem. Since Fuzzy Logic is able to account for uncertainty, vagueness, and non-crisp characteristics, it provides an opportunity to consider the ill-defined components that constitute complexity in Canadas judicial system.

4 Conclusion

In this chapter we have introduced a tool that uses Fuzzy Logic to identify case complexity based on three characteristics that are known before matters come before the court. The number of accused, number of charges, and severity of charges were used to identify complexity. The model should serve as a simplified example of a Fuzzy Logic tool that could be used to identify case complexity. Although only three variables that are known to impact case complexity were employed, there are many characteristics that could be used to predict case complexity. For example, whether or not an accused is likely to plead guilty or elect to hold a trial are factors that may be known before the case enters the courts and are likely important for predicting the length of court processes. Other factors such as whether an accused is in custody at the time the case enters the system are also likely to influence complexity. By including these additional variables into a model, the performance of the model would be expected to be superior to a simplified version.

The model proposed in this chapter is also designed to perform a specific task: predict case complexity based on features of cases known at the early stages of the court process. This type of model, however, could be extended to other applications. For example, case complexity is also an important concept in policing. When police are collecting evidence, making an arrest, or preparing a recommendation for criminal charges, they need to make decisions about the time and resources required to bring the matter to a successful conclusion. If a tool that could predict case complexity early in the investigate stage was available, police could potentially allocate resources more efficiently.

Although the examples presented in this chapter are simplistic, more research needs to be done in this area to determine which parameters are most important in such a case complexity prediction process, and how the membership function for each parameter needs to be positioned. These specific details would have to depend on the actual implementation of the model, and are areas for future consideration.

References

1. BC Court Services Branch: CORIN database, 2004–05 RCC Data, Provincial adult criminal data (2005)
2. Botha, M., von Solms, R.: Utilising fuzzy logic and trend analysis for effective intrusion detection. Comput. Secur. **22**, 423–434 (2003)
3. Brantingham, P.L.: Dynamic modeling of the felony court system. Unpublished PhD Dissertation, Florida State University (1977)
4. Brantingham, P.L , Brantingham, P.J.: Environment, routine, and situation: toward a pattern theory of crime. In: Clarke, R., Felson, M. (eds.) Routine Activity and Rational Choice, Advances in Criminological Theory, vol. 5, (New Brunswick, Transaction Publishers 1993)
5. Canadian Charter of Rights and Freedoms: Part I of The Constitution Act, 1982 being Schedule B to the Canada Act 1982 (U.K.), c. 11 (1982)
6. Dandurand, Y.: Addressing inefficiencies in the criminal justice process. Report prepared for the BC justice efficiencies project criminal justice reform secretariat. International centre for criminal law reform and criminal justice policy. http://www.icclr.law.ubc.ca/files/2009/InefficienciesPreliminaryReport.pdf (2009)
7. Davis, R.C., Smith, B.E., Nickles, L.B.: Specialized Courtrooms: Does Speeding up the Process Jeopardize the Quality of Justice. Criminal Justice Section, American Bar Association, Washington (1996)
8. Friesen, E.C., Geiger, M., Jordan, J., Sulmonetti, A.: Justice in felony courts: a prescription to control delay. Whittier Law Rev. **7**, 15–16 (1979)
9. Goerdt, J.A., Lomvardias, C., Gallas, G., Mahoney, B.: Examining Court Delay: The Pace of Litigation in 26 Urban Trial Courts. National Center for State Courts, Williamsburg (1989)
10. Grubesic, T.: On the application of fuzzy clustering for crime hot spot detection. J. Quant. Criminol. **22**, 77–105 (2006)
11. Hall, N.: B.C'.s top judges 'gravely concerned' about court delays caused by sheriff shortage. The Vancouver sun. http://www.vancouversun.com/sports/judges+gravely+concerned+about+court+delays+caused+sheriff+shortage/4960898/story.html. Accessed 16 June 2011
12. Office, Home: Adjournments in Magistrates Courts. Home Office Research and Statistics Department, London (1990)
13. Justice Management Institute: Improving your jurisdictions felony caseflow process: A primer on conducting an assessment and developing an action plan. Prepared by the Justice Management Institute for the Bureau of Justice Assistance, Criminal Courts Technical Assistance Project, American University (2000)
14. Kelly, M.J., Levy, E.: Making the System Work in the Baltimore Criminal Justice System: An Evaluation of Early Disposition Court. Philadelphia, Center for Applied Research (2002)
15. Lamer, A.: The role of judges. In: Campion, J., Badovinac, E. (eds.) The empire club of Canada speeches 1994–1995 (Toronto: The Empire Club Foundation, 1995), pp. 124–136. http://www.empireclub foundation.com (1995)
16. LeSage, P.J., Code, M.: Report of the review of large and complex criminal case procedures. Report submitted to the Honourable Chris Bentley, Attorney General of Ontario. Ontario Ministry of the Attorney General (2008)
17. Li, S., Kuo, S., Tsai, F.: An intelligent decision-support model using FSOM and rule extraction for crime prevention. Expert Syst. Appl. **37**, 7108–7119 (2010)
18. Liao, N., Tian, S., Wang, T.: Network forensics based on fuzzy logic and expert system. Comput. Commun. **32**, 1881–1892 (2009)
19. Zadeh, Lotfi A.: Fuzzy sets. Inf. Control **8**(3), 338–353 (June 1965)
20. Mago, V.K., Frank, R., Reid A.A., Dabbaghian, V. The strongest does not attract all but it does attract the most–Evaluating the criminal attractiveness of shopping malls using fuzzy logic. Expert systems (2013) doi: 10.1111/exsy.12015

21. McLachlin, B. (P.C.): The challenges we face. Remarks of the Right Honourable Beverley McLachlin, P.C., Presented at the Empire Club of Canada, Toronto. http://www.scc-csc.gc.ca/court-cour/ju/spe-dis/bm07-03-08-eng.asp. Accessed 8 March 2007
22. Mohajerani, M., Moeini, A., Kianie, M.: NFIDS: a neuro-fuzzy intrusion detection system. In: Proceedings of the 2003 10th IEEE International Conference on Electronics, Circuits and Systems (ICECS 03), vol. 1, pp. 348351 (2003)
23. Neubauer, D.W., Lipetz, M.J., Luskin, M.L., Ryan, J.P.: Managing the Pace of Justice: An Evaluation of LEAA'S Court Delay-Reduction Programs. U.S. Department of Justice, Washington (1981)
24. Passino, K.M., Yurkovich, S.: Fuzzy Control. Addison Wesley Longman, Menlo Park (1998)
25. Payne, J.: Criminal Trial Delays in Australia: Trial Listings Outcomes. Canberra, Australian Institute of Criminology (2007)
26. Askov, R. v.: 2 S.C.R. 1199 (1990)
27. Scheb, J.M., Scheb, J.M.: (II), : Criminal Procedure, 6th edn. Wadsworth Cengage Learning, Belmont (2009)
28. Sipes, L.L., Carlson, A.M., Tan, T., Aikman, A.B.: Page. Managing to reduce delay. National Centre for State Courts, R.W. (1980)
29. Statistics Canada: Measuring crime in Canada: Introducing the crime severity index and improvements to the uniform crime reporting survey. http://www.statcan.gc.ca/pub/85-004-x/85-004-x2009001-eng.htm (2009)
30. Steering Committee on Justice Efficiencies and Access to the Justice System: The final report on early case consideration of the steering committee on justice efficiencies and access to the justice system. Department of Justice Canada. http://www.justice.gc.ca/eng/esc-cde/ecc-epd.pdf (2006)
31. The Provincial Court of British Columbia: Justice delayed: A report of the Provincial Court of British Columbia concerning judicial resources, 14 Sept 2010
32. Thomas, J.: Adult criminal court statistics, 2008/2009: Case processing in adult criminal courts. Canadian Centre for Justice Statistics. http://www.statcan.gc.ca/pub/85-002-x/2010002/article/11293-eng.htm (2010)
33. Verma, A.: Construction of offender profiles using fuzzy logic. Policing Int. J. Police Strateg. Manag. **20**, 408–418 (1997)
34. Whittaker, C., Mackie, A. with Lewis, R., Ponikiewski, N.: Managing courts effectively: the reasons for adjournments in Magistrates courts. Home office research study No. 168. London: Home office research and statistics department (1997)

Chapter 8
Understanding the Impact of Face Mask Usage Through Epidemic Simulation of Large Social Networks

Susan M. Mniszewski, Sara Y. Del Valle, Reid Priedhorsky, James M. Hyman and Kyle S. Hickman

Abstract Evidence from the 2003 SARS epidemic and 2009 H1N1 pandemic shows that face masks can be an effective non-pharmaceutical intervention in minimizing the spread of airborne viruses. Recent studies have shown that using face masks is correlated to an individual's age and gender, where females and older adults are more likely to wear a mask than males or youths. There are only a few studies quantifying the impact of using face masks to slow the spread of an epidemic at the population level, and even fewer studies that model their impact in a population where the use of face masks depends upon the age and gender of the population. We use a state-of-the-art agent-based simulation to model the use of face masks and quantify their impact on three levels of an influenza epidemic and compare different mitigation scenarios. These scenarios involve changing the demographics of mask usage, the adoption of mask usage in relation to a perceived threat level, and the combination of masks with other non-pharmaceutical interventions such as hand washing and social distancing. Our results shows that face masks alone have limited impact on the spread of influenza. However, when face masks are combined with other interventions such

S. M. Mniszewski (✉) · S. Y. Del Valle · R. Priedhorsky
Los Alamos National Laboratory, Los Alamos, NM 87545, USA
e-mail: smm@lanl.gov

S. Y. Del Valle
e-mail: sdelvall@lanl.gov

R. Priedhorsky
e-mail: reidpr@lanl.gov

J. M. Hyman and K. S. Hickman
Tulane University, New Orleans, LA 70118, USA
e-mail: myhyman@tulane.edu

K. S. Hickman
e-mail: khickman@tulane.edu

V. Dabbaghian and V. K. Mago (eds.), *Theories and Simulations of Complex Social Systems*, Intelligent Systems Reference Library 52,
DOI: 10.1007/978-3-642-39149-1_8, © Springer-Verlag Berlin Heidelberg 2014

as hand sanitizer, they can be more effective. We also observe that monitoring social internet systems can be a useful technique to measure compliance. We conclude that educating the public on the effectiveness of masks to increase compliance can reduce morbidity and mortality.

1 Introduction

Pharmaceutical interventions such as vaccines and antiviral medication are the best defense in reducing morbidity and mortality during an influenza pandemic. However, current egg-based vaccine production process can take up to 6 months for the development and availability of a strain-specific vaccine and antiviral supplies may be limited. Fortunately, alternative strategies such as non-pharmaceutical interventions can reduce the spread of influenza until a vaccine becomes available. Face masks have been used to combat airborne viruses such as the 1918–1919 pandemic influenza [4, 29], the 2003 SARS outbreak [7, 38], and the most recent 2009 H1N1 pandemic [12]. These studies indicate that if face masks are readily available, then they may be more cost-effective than other non-pharmaceutical interventions such as school and/or business closures [13].

We focus on the use of surgical face masks and N95 respirators (also referred to as face masks). A surgical mask is a loose-fitting, disposable device that prevents the release of potential contaminants from the user into their immediate environment [8, 40]. They are designed primarily to prevent disease transmission to others, but can also be used to prevent the wearer from becoming infected. If worn properly, a surgical mask can help block large-particle droplets, splashes, sprays, or splatter that may contain germs (viruses and bacteria), and may also help reduce exposure of saliva and respiratory secretions to others. By design, they do not filter or block very small particles in the air that may be transmitted by coughs or sneezes.

An N95 respirator is a protective face mask designed to achieve a very close facial fit and efficient filtration of airborne particles [40]. N95 respirators are designed to reduce an individual's exposure to airborne contaminants, such as infectious viral and bacterial particles, but they are also used to prevent disease transmission when worn by a sick individual [20]. Typically, they are not as comfortable to use as a surgical face mask, and some health care workers have found them difficult to tolerate [23]. N95 respirators are designed for adults, not for children, and this limits their use in the general population.

Surgical masks and N95 respirators have been found to be equally effective in preventing the spread of influenza in a laboratory setting [20] as well as for health care workers [24]. In addition to reducing the direct flow of an airborne pathogen into the respiratory system the masks act as a barrier between a person's hands and face, which can reduce direct transmission.

A survey paper by Bish and Michie [5] on demographic determinants of protective behavior showed that compliance to using face masks is tied to age and gender. They observed that females and older adults were more likely to accept protective behaviors than other population groups. Supporting these ideas, usage of face masks was consistently higher among females than male metro passengers in Mexico City during the 2009 Influenza A (H1N1) pandemic [12]. Limited studies suggest that there is more social stigmatization associated with wearing face masks in Western Countries than in Asia. For example, people rarely wear face masks in public in the United States, compared with their use in Japan and China [17]. An article published in 2009 by *New York Times Health* reported that "masks scare people away from one another" resulting in an unintentional social distancing measure [30] or "stay away" factor. Pang et al. showed that during the 2003 SARS outbreak, non-pharmaceutical interventions where implemented followed the epidemic curve [33]. That is, as the perception of SARS increased, more measures were implemented, and as the incidence declined, several measures were relaxed.

Based on these studies, we investigate the impact of face mask usage on the spread of influenza under several assumptions, including: (1) that females and older people will be more likely to wear them, (2) face mask wearers may follow the epidemic (e.g., the number of people wearing face masks depends on the incidence), and (3) face masks scare people away.

In order to transfer our results to the real world, it will be important to measure compliance. In the case of interventions such as face mask use, where individuals often choose to comply or not comply in the privacy of their daily lives, traditional methods of measuring compliance may be ineffective. Accordingly, we turn to social internet systems, specifically Twitter, where users share short text messages called *tweets*. These messages are directed to varying audiences but are generally available to the public regardless; they are used to share feelings, interests, observations, desires, concerns, and the general chatter of daily life. While other researchers have used Twitter to measure public interest in various health topics, including face masks as an influenza intervention [35], we carry out a brief experiment to explore the feasibility of using tweets to measure *behavior*.

The goal of this study is to understand the effectiveness of face mask usage for influenza epidemics of varying strengths (high, medium, low). A high level epidemic would be similar to the 1918–1919 H1N1 "Spanish flu" outbreak with large morbidity and mortality [32, 34, 42], a medium level would be similar to the 1957–1958 H2N2 Asian flu [15, 18], and a low level would be similar to the more recent 2009 Novel H1N1 flu [6, 10, 19]. We simulate face mask usage behavior through detailed large-scale agent-based simulations of social networks. These simulations have been performed using the Epidemic Simulation System (EpiSimS) [27, 28, 37] described in the next section.

2 Methods

2.1 Agent-Based Model Description

EpiSimS is an agent-based model that combines three different sets of information to simulate disease spread within a city:

- population (e.g., demographics),
- locations (e.g., building type and location), and
- movement of individuals between locations (e.g., itineraries).

We simulated the spread of an influenza epidemic in southern California with a synthetic population constructed to statistically match the 2000 population demographics of southern California at the census tract level. The synthetic population consists of 20 million individuals living in 6 million households, with an additional 1 million locations representing actual schools, businesses, shops, or social recreation addresses. The synthetic population of southern California represents only individuals reported as household residents in the 2000 U.S. Census; therefore, the simulation ignores visiting tourists and does not explicitly treat guests in hotels or travelers in airports.

We use the National Household Transportation Survey (NHTS) [44] to assign a schedule of activities to each individual in the simulation. Each individual's schedule specifies the starting and ending time, the type, and the location of each assigned activity. Information about the time, duration, and location of activities is obtained from the NHTS. There are five types of activities: *home, work, shopping, social recreation,* and *school,* plus a sixth activity designated *other.* The time, duration, and location of activities determines which individuals are together at the same location at the same time, which is relevant for airborne transmission.

Each location is geographically-located using the Dun and Bradstreet commercial database and each building is subdivided based on the number of activities available at that location. Each building is further subdivided into rooms or mixing places. Schools have classrooms, work places have workrooms, and shopping malls have shops. Typical room sizes can be specified; for example, for workplaces, the mean workgroup size varies by standard industry classification (SIC) code. The number of sub-locations at each location is computed by dividing the location's peak occupancy by the appropriate mixing group size. We used two data sources to estimate the mean workgroup by SIC, including a study on employment density [45] and a study on commercial building usage from the Department of Energy [26]. The mean workgroup size was computed as the average from the two data sources (normalizing the worker density data) and ranges from 3.1 people for transportation workers to 25.4 for health service workers. The average over all types of work is 15.3 workers per workgroup. For the analyses presented here, the average mixing group sizes are: 8.5 people at a school, 4.4 at a shop, and 3.5 at a social recreation venue.

2.2 Disease Progression Model

Airborne diseases spread primarily from person-to-person during close proximity through contact, sneezing, coughing, or via fomites. In EpiSimS, an interaction between two individuals is represented only by:

- when they begin to occupy a mixing location together,
- how long they co-occupy within a mixing place,
- a high-level description of the activity they are engaged in, and
- the ages of the two individuals.

A location represents a street address, and a room or mixing place represents a lower-level place where people have face-to-face interactions. When an infectious person is in a mixing location with a susceptible person for some time, we estimate a probability of disease transmission, which depends on the last three variables listed above. Details of social interactions such as breathing, ventilation, fomites, moving around within a sub-location, coughing, sneezing, and conversation are not included. Disease transmission between patients and medical personnel is not handled explicitly, and no transmission occurs when traveling between activities. Note that individuals follow a static itinerary, except when they are sick or need to care for a sick child. In this case, their schedule changes and all activities they were supposed to undertake are changed to *home*.

If susceptible person j has a dimensionless susceptibility multiplier S_j, infectious person I has an infectious multiplier I_i and T is the average transmissibility per unit time, then, $T S_j I_i$ will be the mean number of transmission events per unit time between fully infectious and fully susceptible people. The sum

$$\sum_j T S_j I_i$$

extends over all infectious persons that co-occupied the room with individual j. For events that occur randomly in time, the number of occurrences in a period of time of length t obeys a Poisson probability law with parameter.

$$\sum_j T S_j I_i t$$

Thus, the probability of no occurrences in time interval t is

$$e^{-\sum_j T S_j I_i t}$$

and the probability of at least one occurrence is

$$1 - e^{-\sum_j T S_j I_i t}$$

Using the mean duration t_{ij} of contacts between a susceptible person j and infectious person i, we assume that the probability that susceptible individual j gets infected during an activity is computed as:

$$P_j = 1 - e^{-\sum_j TS_j I_i t_{ij}} \tag{1}$$

Disease progression is modeled as a Markov chain consisting of five main epidemiological stages: uninfected, latent (non-infectious), incubation (partially infectious), symptomatic (infectious), and recovered. The incubation and symptomatic stage sojourn time distributions are described by a half-day histogram, giving respectively the fraction of cases that incubate for a period of between 0 and 0.5 days, 0.5 and 1.0 days, etc., before transitioning to the symptomatic or recovered stages, respectively. The average incubation time is 1.9 days and average duration of symptoms is 4.1 [25]. The influenza model assumes that 50% of adults and seniors, 75% of students, and 80% of pre-schoolers will stay at home soon within 12 hrs of the onset of influenza symptoms. These people can then transmit disease only to household members or visitors. In addition, based on previous studies [25], we assume that 33.3% of infections are subclinical where an infected individual is asymptomatic and shows no sign of infection. We modeled the subclinical manifestation as only half as infectious as the symptomatic manifestations. Persons with subclinical manifestations continue their normal activities as if they were not infected. The assumed hospitalization rate is a percentage of symptomatic individuals dependent on the strength of the pandemic. To simulate the higher attack rates seen in children, we assume that the infection rate in children was double that in adults. We analyze multiple scenarios for the same set of transmission parameters where the population was initially seeded with 100 people infected, all in the incubation stage.

2.3 Behavior Model

The behavior of each individual (agent) in an EpiSimS simulation is defined based on distributions for the effectiveness of their face mask usage in preventing infection to others (given as a distribution), effectiveness to preventing the individual from becoming infected (given as a distribution), acceptance of using the mask (given as a distribution), along with applicable age range, gender, and other possible demographic descriptive information. Effectiveness to others for mask usage is based on the protection factor of a mask type. It is the protection provided to people in contact with a sick individual wearing a mask. Effectiveness to self is based on the penetration level of a mask type. It is the protection provided to a healthy individual when in close contact with an infectious person. Distributions were used based on mask testing for the penetration level [2, 9, 21, 31] and protection factor [22]. Examples of these distributions are shown for N95 respirators in Table 1 and for surgical masks in Table 2. The effectiveness values drawn from each distribution are used to modify

Table 1 Effectiveness of N95 respirators in preventing an infected person from infecting others (protection factor) and the effectiveness of the face mask to prevent the wearer from being infected (penetration level) are listed along with the percentage of face mask users with this level of effectiveness from testing

Effectiveness to others (protection factor)	N95 respirator (%) users	Effectiveness to self (penetration level)	N95 (%) users
less than 0.1	0.00	less than 0.5	9.52
0.1	87.88	0.5	9.52
0.5	12.12	0.6	14.29
		0.7	14.29
		0.8	33.33
		0.9	19.05

Table 2 Effectiveness of surgical masks in preventing an infected person from infecting others (protection factor) and the effectiveness of the face mask to prevent the wearer from being infected (penetration level) are listed along with the percentage of face mask users with this level of effectiveness from testing

Effectiveness to others (protection factor)	Surgical mask (%) users	Effectiveness to self (penetration level)	Surgical mask (%) users
<0.1	91.67	0.1	13.89
0.1	8.33	0.2	8.33
		0.3	5.55
		0.5	5.55
		0.6	11.12
		0.7	38.89
		0.8	16.67

the infectivity (I_i) and susceptibility (S_j) between pairs contributing to whether or not transmission occurs.

As stated previously, age and gender play an important role in determining whether someone will comply with wearing a mask. The age ranges and compliance or acceptance by gender are based on values from a survey of behavior studies [5] and are shown in Table 3. Simulations that assigned mask usage by age and gender used the age ranges and acceptance in this table. Simulations that assigned mask usage randomly used constant acceptance values (e.g., 25 % of the population) for adults-only or all.

We assume that willingness to wear a mask is not influenced by a person being ill and the masks are only worn in non-home settings. Mask usage is initiated as an exogenous event, specified for a range of days. Usage can be specified as a fraction of all possible users (based on age and gender) and the duration can be specified as a distribution (e.g., constant, normal). Early in the simulations, each individual determines whether they will wear a mask based on age, gender, and acceptance. This is the pool of people from which mask users are selected. When we assume that mask usage will follow the course of an epidemic (e.g., disease perception increases

Table 3 Face mask
acceptance by gender and
age. Notice that the
willingness to use a face mask
increases with age and that
women are more willing to
use a face mask than men of
the same age

Age group	Males (%)	Females (%)
6–15	33	33
16–24	33	54
25–34	45	63
35–44	59	74
45–54	55	68
55–64	59	71
65–74	63	75
75+	57	72
Average	57	64

as incidence increases and vice-versa), mask usage ramps up and then down. For this
scenario, mask users change over time and some may use masks for a sequence of
days multiple times.

Scenarios that take into account a stay away factor used higher effectiveness
values based on assumptions regarding the amount of social distancing we expect
a mask wearer to experience (e.g., 30 %). The mechanism we are assuming here is
that, in general, individuals will attempt to limit their contact with a person wearing
a mask. This translates to a larger histogram bin size for the distribution. Scenarios
where both surgical masks and hand sanitizer served as the mitigation strategy, do not
use the protection level and penetration factor values for effectiveness as described
previously, instead an effectiveness value of 50 % is used based on an intervention
trial conducted at the University of Michigan [1].

2.4 The Reproduction Number

In epidemiological models, the effectiveness of mitigation strategies are often measured by their ability to reduce the effective reproduction number or replacement
number R_{eff}. R_{eff} is the average number of secondary cases produced by a typical
infectious individual during their infectious period [46]. In a completely susceptible
population and in the absence of mitigation strategies, the average number of secondary cases is referred to as R_0. The magnitude of R_0 determines whether or not
an epidemic will occur and if so, its severity. The number of infections grows when
R_0 is greater than one and it dies out when R_0 is less than one.

3 Results

We compare a base case scenario where no face masks are used for the high, medium,
and low epidemic levels with simulations using only face masks, face masks and hand
sanitizer (M and HS), and face masks coupled with social distancing (M and SD).

For the base case scenarios, we compare the epidemic parameters related to morbidity and mortality, including the attack rate, clinical attack rate, hospitalization rate, and mortality rate.

All of the scenarios that include face mask usage mitigations allow mask base acceptance by age and gender. Additionally, mask users follow the course of the epidemic incidence, increasing to the peak and then decreasing, ending 4 weeks after the peak. In support of this behavior, we present the results of a small experiment, where we use Twitter to estimate the shape of the compliance curve with respect to face masks.

Surgical masks and N95 respirators are considered independently in the face mask only scenarios, while surgical masks are the choice for the hand sanitizer and social distancing scenarios. N95 respirators can be more effective if both adults and children would use them, but they have not been designed for children and can be uncomfortable even for adults for long-term use. For these scenarios where mitigations are implemented, we compare the clinical attack rate, effective reproductive number, and for some cases, we show the the disease prevalence (symptomatic cases), incidence of mask users (new cases), and the effective reproductive number over time (R_{eff}).

3.1 Base Case Scenario

As described earlier, we used influenza epidemics of varying strengths (high, medium, low) to compare the impact of face mask usage on controlling the spread. These different levels share a similar disease progression as described in Sect. 2. The high level epidemic is based on the 1918–1919 H1N1 "Spanish flu" outbreak and has large morbidity and mortality [32, 34, 42], the medium level is based on the 1957–1958 H2N2 Asian flu [15, 18], and the low level is based on the more recent 2009 Novel H1N1 flu [6, 10, 19]. The number of hospitalizations and deaths were extrapolated from the U.S. population during the represented pandemic year to the U.S. synthetic population of 280M (based on 2000 census data). The attack rate (percentage of population infected), clinical attack rate (percentage of population symptomatic), hospital rate (hospitalizations out of population), and mortality rate (deaths out of population) are shown for each strength in Table 4. Figure 1 shows each of their respective epidemic curves for the new symptomatic as a function of time.

Table 4 Epidemic parameters associated with high, medium, and low strengths of epidemic

Epidemic level	Attack rate (%)	Clinical attack rate (%)	Hospital rate (%)	Mortality rate (%)
High	40.0	30.0	0.500	0.300
Medium	30.0	19.7	0.250	0.100
Low	20.0	10.0	0.008	0.015

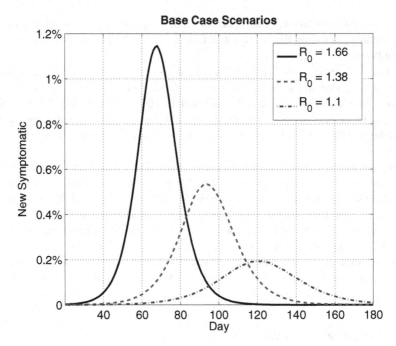

Fig. 1 Base case simulation results for the three different epidemic strengths, showing the percentage of the population that becomes symptomatic per day

3.2 Using Twitter to Quantify Face Mask Usage

Our goal in exploring Twitter is to evaluate two conjectures: first, that the level of face mask wearing follows the disease incidence level, and second, that analysis of the public tweet stream is a feasible technique to measure compliance with face mask wearing (and, by implication, other behaviors relevant to infectious disease). To do so, we analyzed tweets published globally between September 6, 2009 and May 1, 2010, roughly corresponding to the H1N1 pandemic flu season in the United States.

There are 548,893,258 tweets in our dataset, an approximate 10 % sample of total Twitter traffic during this period. Of these, we selected the 75,946 which contained the word "mask"; in turn, a small fraction of these keyword matches—we estimate 3,350, or about 4.5 %—actually concern the medical face masks of interest to the present work (topics also include costume, sports, metaphor, cosmetics, movies, and others).

In order to identify these relevant tweets, we manually examined a random sample of 7,602 keyword matches (roughly 10 % of the total), coding them as (a) mentioning medical face masks (335 tweets), and perhaps additionally (b) sharing a specific observation that either the speaker or someone else is wearing, or has recently worn, a face mask (138 tweets).

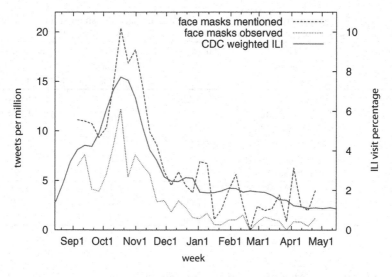

Fig. 2 Of each million tweets during the period September 6, 2009 through May 6, 2010, we show the number in which face masks are mentioned, as well as the subset of mentions which observe that someone specific is or was recently wearing a mask, whether the speaker himself or someone else. Also shown is the influenza-like illness rate from the CDC for the same period. The Pearson correlation between ILI rate and mentions is 0.92, and between ILI rate and observations is 0.90

Our results are shown in Fig. 2. As noted above, there are very limited survey studies that have collected information on mask use, especially from Western Countries [5]; accordingly, we compare our Twitter mention and observation counts against influenza-like illness (ILI) data published by the Centers for Disease Control (CDC) [11]. The correlation is excellent: 0.92 for mentions and 0.90 for observations.

These results have two implications. They provide empirical support for our assumption that face mask use is disease-dependent; that is, as disease incidence increases, face mask use increases, and as incidence decreases, so does mask use. Also, they suggest more broadly that social internet systems such as Twitter can, in fact, be used to measure disease-relevant behavior in the real world.

Challenges remain, however. First, we point out the severe signal-to-noise of these data: we identified just 20 out of every million tweets as relevant, even at the peak of the epidemic. Accordingly, analysis focusing on specific locales or demographic groups is not possible with this approach. Second, our manual coding approach clearly does not scale. Finally, we strongly suspect that information relevant to our specific questions (e.g., How many people are using face masks? Who are they? Where are they?) is contained in the vast number of tweets our coarse, preliminary approach discards as irrelevant. Our future work in measuring real-world behavior will go beyond simple keyword searches to leverage more sophisticated data mining algorithms.

Table 5 Attack rate parameters associated with high, medium, and low strengths of epidemic for face mask only scenarios starting when 0.01 % of the population is symptomatic

Epidemic level	Mask scenario	Attack rate (%)	Overall Clinical attack rate (%)	Mask users Clinical attack rate (%)
High	Surgical mask	34.22	25.66	14.24
	N95 respirator adults	35.03	26.27	12.74
	N95 respirator all	32.26	24.20	12.09
Medium	Surgical mask	24.51	16.35	7.40
	N95 respirator adults	25.55	17.04	7.03
	N95 respirator all	23.40	15.60	5.89
Low	Surgical Mask	16.35	8.18	2.88
	N95 Respirator Adults	17.69	8.85	2.80
	N95 Respirator All	16.96	8.49	1.73

3.3 Comparison of Intervention Strategies

Face mask only mitigation strategies were considered for surgical masks and N95 respirators separately. All scenarios began when 0.01 or 1.0 % of the population was symptomatic. Usage was based on age and gender and followed the course of the epidemic. Surgical masks were available to all age groups and N95 respirators to adults only and all age groups. Since N95 respirators were not designed for use by children, the adults only scenario is more realistic; however the all age groups scenario allows us to understand the importance of children wearing masks and the use of a more protective mask.

Scenarios with face mask usage starting when 1.0 % of the population was symptomatic resulted in higher attack rates and clinical attack rates than that for 0.01 % and will not be considered further here. Those starting at 0.01 % slowed the epidemic, allowing less burden to the public health system.

Table 5 shows the overall clinical attack rates for the epidemic as well as just for the mask users for all scenarios and epidemic strengths. Overall, only a small improvement is seen over the base case. The maximum mask users for all scenarios is 45–50 % of the population. Considering only the mask users, the clinical attack rates are much improved, with significant reductions for all three scenarios. The largest improvement is seen for N95 respirator where use is not limited to adults. This shows the importance of involving children in a face mask mitigation. Of the more realistic scenarios, surgical mask and N95 respirator adults, surgical mask performs best overall for all pandemic strengths, though worst when only considering mask users.

We compare the impact of combining face masks with hand sanitizers (M and HS) or with social distancing (M and SD). As described in Sect. 2.3, M and HS are assumed to reduce the transmission rate by 50 % and M and SD are assumed to reduced the transmission rate by 30 %. Figure 3, part A and C shows the epidemic curves when M and HS are implemented after 1.0 % of the population is symptomatic, and M and SD

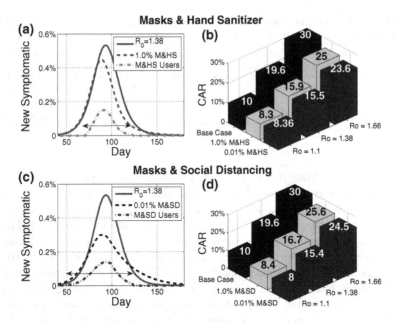

Fig. 3 Results of surgical masks and hand sanitizers (*top*) and masks and social distancing (*bottom*). **a** Epidemic curves for the base case, when the intervention is implemented after 1.0 % of the population is symptomatic, and the population that adopts the behavior (M and HS users). **b** Clinical attack rates (CAR) for the various pandemic levels and when masks and hand sanitizers are implemented after 1.0 and 0.01 % of the population is symptomatic. **c** Epidemic curves for the base case, when the intervention is implemented after 0.01 % of the population is symptomatic, and the population that adopts the behavior (M and SD users). **d** Clinical attack rates (CAR) for the base case, and two mask and social distancing scenarios for the different pandemic levels

when 0.01 % of the population is symptomatic, respectively. In addition to showing the overall dynamics of these two interventions, we show the epidemic curve for individuals who adopted the specified behavior, but who still became infected. Note that although the clinical attack rate was only reduced by 19 and 21 % for these two scenarios, the clinical attack rate for M and HS users was only 3.6 or an 81 % reduction. Similarly, the clinical attack rate for the M and SD users is 4.7 or a 76 % reduction from the base case. Part B and D, shows the clinical attack rate for various assumptions of the M and HS and M and SD scenarios and all the different pandemic levels.

From the results, it is clear that the earlier the interventions are put in place, the higher the impact they will have on reducing morbidity and mortality. Although these non-pharmaceutical interventions may not be very effective when compared to vaccines and antivirals, the overall impact for people that adopt these behaviors is significantly lower than the epidemic curve for the entire population. Table 6 takes the new clinical attack rate for the M and HS and M and SD intervention strategies and computes their difference. Then, this difference is expressed in the table as a percentage of the base case clinical attack rate for that epidemic strength. This

Table 6 Difference in clinical attack rate as a percent of base case clinical attack rate when comparing M and SD and M and HS intervention strategies

R_0	0.01 M and HS (%) 0.01 M and SD (%)	1.00 M and HS (%) 1.00 M and SD (%)	1.00 M and HS (%) 0.01 M and SD (%)	0.01 M and HS (%) 1.00 M and SD (%)
1.10	3.60	1.00	3.00	0.40
1.38	0.51	4.10	2.60	6.12
1.66	3.00	2.00	1.70	6.70

is meant to demonstrate the difference in the clinical attack rate relative to each intervention strategy on a scale that is proportional to the base case. If this percent is small then one could reasonably conclude that there is not much difference in the intervention strategies at that level. Overall, the scenarios with masks and hand sanitizer had a difference of less than 10 % of the base case clinical attack rate in all cases (see Table 7). The case of comparing M and HS implemented when 0.01 % of the population is symptomatic and M and SD when 1.0 % of the population is symptomatic is especially interesting at a low epidemic level, since the difference is less than 5 % even though M and SD has only a 30 % effectiveness compared to M and HS 50 % effectiveness. This motivates future studies into the difference in the effectiveness of these two intervention strategies at various epidemic strengths.

To better understand the overall effectiveness of the different intervention strategies we compare the effective reproduction number, R_{eff}, for five different scenarios:

- Surgical mask only (Mask),
- N95 respirators only-adults (N95 Adult),
- N95 respirators only-all (N95 All),
- Surgical masks and social distancing (Mask and Social Distancing), and
- Surgical masks and hand sanitizer (Mask and Hand Sanitizer).

All scenarios assume that the intervention begins when 0.01 % of the population is symptomatic, follows the course of the epidemic (ramping up to the peak and then down), and lasts 4 weeks after the peak. The likelihood of use of a non-pharmaceutical intervention, in each scenario, was dependent on age and gender as discussed previously.

Table 7 Percent reduction in clinical attack rate from base case at different epidemic strengths for M and HS or M and SD implemented at different epidemic levels

R_0	M and HS		M and SD	
	0.01 (%)	1.00 (%)	0.01 (%)	1.00 (%)
1.10	16.40	17.00	20.00	16.00
1.38	20.90	18.90	21.40	14.80
1.66	21.30	16.70	18.30	14.67

Note that at low epidemic levels, if implemented early, social distancing is competitive with hand sanitizing as an intervention strategy

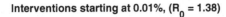

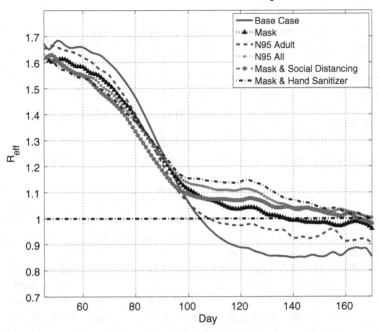

Fig. 4 R_{eff} over time as the epidemic progresses. For five different scenarios (shown starting from day 40), the dynamic behavior of R_{eff} is different. Intervention strategies cause the initial R_{eff} to be smaller than the base case, and then take longer to decrease below $R_{\text{eff}} = 1$. (The N95 Adult case has an initially higher R_{eff} than the other scenarios, presumably since children did not have intervention in this case.)

Figure 4 shows the change in the effective reproduction number, R_{eff}, over the course of the epidemic for the five scenarios described above during a medium ($R_0 = 1.38$) level outbreak. The basic reproduction number, R_0, is the average number of cases generated by a typical infectious individual in a completely susceptible population. Similarly, the effective reproduction number is the average number of cases generated by an infectious individual in a population that is not completely susceptible. The magnitude of the reproduction number determines whether or not an epidemic occurs and what its severity will be. When $R_0 > 1$, the number of infections grow and an epidemic occurs, and when $R_0 < 1$, the epidemic goes extinct.

We notice (Fig. 4) that for the different intervention strategies, the maximum R_{eff} is reduced. The exception is for the N95 scenario, N95 Adult, when children do not wear masks. In this case, R_{eff} shows a dramatic decrease but starts out high; this exception is not present if children wear the respirators as in N95 All.

4 Discussion

Non-pharmaceutical interventions such as face masks can play an important role in controlling the spread of airborne viruses. Based on historical observations, it is clear that some people wear face masks to protect themselves from infection. However, due to their limited effectiveness (known from filtration performance tests) the impact of face masks at the population level has not been well studied.

We used an agent-based simulation model to examine the effect that face masks alone, and in combination with other non-pharmaceutical interventions, has on reducing the spread of influenza. We analyzed the sensitivity with respect to various parameters including pandemic level, type of face mask, timing of intervention(s), and type of intervention.

Our results show that, in general, face masks have an impact on reducing the overall incidence and extending the length of the epidemic. Masks alone reduce the clinical attack rate, on average, by over 10 % for the entire population and 50 % for the population that wears face masks. Not surprisingly, our results show that face masks are more effective when coupled with other interventions. Although we expected that masks and hand sanitizers would have the largest return (given that we assume to be 50 % effective), social distancing performed almost as well as the hand sanitizer (even though we assume it was only 30 % effective). These observations imply that any mitigation that aims at reducing the probability of transmission, regardless of effectiveness, can contribute in reducing the overall impact of disease. Furthermore, the results are consistent with other studies concluding that the earlier interventions are put in place, the higher the impact they have on reducing morbidity and mortality.

We compare the effective reproduction numbers for various scenarios and show that intervention strategies cause the initial R_{eff} to be smaller than the base case and take longer to decrease below $R_{eff} = 1$. We also noted that the N95 case had an initially higher R_{eff} than the other scenarios due to the assumption that children would not wear N95 respirators.

For any intervention, it is important to measure the rate at which the intervention is actually happening. Non-pharmaceutical interventions such as face mask wearing presents special problems in this regard, because the decision to comply or not comply is an individual one which takes place away from observation by health providers. The intuition in exploring social internet systems such as Twitter to make these measurements is that the very high volume of observations, perceptions, and desires can, in aggregate, provide a sufficiently accurate measurement of compliance in real-world settings. Our preliminary results in analyzing Twitter are consistent with this intuition: we measured the use of face masks with a simple keyword-based approach, and both mentions of and observations of wearing face masks correlate strongly with CDC influenza incidence data. We expect future efforts to deepen this capability, providing results segmented by locale or demographics.

We conclude that for mathematical models of infectious diseases to be useful in guiding public health policy, they need to consider the impact of non-pharmaceutical interventions. Face masks can be a cost-effective intervention when compared to closures; therefore, public health campaigns should focus on increasing compliance. Additionally, measuring the effect of these campaigns should include analysis of social internet systems and other emerging data sources. The results presented here are useful in providing estimates of the effects of non-pharmaceutical interventions on the spread of influenza.

Acknowledgments We would like to acknowledge the Institutional Computing Program at Los Alamos National Laboratory for use of their HPC cluster resources. We thank Aron Culotta for his assistance with the Twitter data analysis. We also thank Geoffrey Fairchild for providing some useful articles. This research has been supported at Los Alamos National Laboratory under the Department of Energy contract DE-AC52-06NA25396 and a grant from NIH/NIGMS in the Models of Infectious Disease Agent Study (MIDAS) program (U01-GM097658-01).

References

1. Aiello, A.E., Perez, V., Coulborn, R.M., et al.: Facemasks, hand hygiene, and influenza among young adults: a randomized intervention trial. PLoS One **7**(1), e29744 (2012)
2. Balazy, A., Toivola, M., Adhikari, A. et al.: Do N95 respirators provide 95% protection level against airborne virus, and how adequate are surgical masks? Am. J. Infect. Control **34**(2), 51–57 (2006)
3. Barr, M., Raphael, B., Taylor, M. et al.: Pandemic influenza in Australia: using telephone surveys to measure perceptions of threat and willingness to comply. BMC Infect. Dis. **8**, 117 (2008)
4. Billings, M.: The influenza pandemic of 1918: the public health response. http://virus.stanford.edu/uda/fluresponse.html (2005). Accessed 26 April 2012
5. Bish, A., Michie, S.: Demographic and attitudinal determinants of protective behaviours during a pandemic: a review. Br. J. Health Psych. **15**, 797–824 (2010)
6. Bronze, M.S.: H1N1 influenza (swine flu). Medscape reference. http://emedicine.medscape.com/article/1807048-overview (2012). Accessed 27 April 2012
7. Brookes, T., Khan, O.A.: Behind the mask: how the world survived SARS, the first epidemic of the twenty-first century. American Public Health Association, Washington, DC (2005)
8. Brosseau, L., Ann, R.B.: N95 respirators and surgical masks. http://blogs.cdc.gov/niosh-science-blog/2009/10/n95/ (2012). Accessed 11 May 2012
9. Centers for Disease Control and Prevention: Laboratory performance evaluation of N95 filtering respirators, 1996. http://www.cdc.gov/mmwr/preview/mmwrhtml/00055954.htm#00003611.htm (1998). Accessed 26 April 2012
10. Centers for Disease Control and Prevention: CDC estimates of 2009 H1N1 influenza cases, hospitalizations and deaths in the United States, April 2009–January 16, 2010. http://www.cdc.gov/h1n1flu/estimates/April_January_16.htm (2010). Accessed 27 April 2012
11. Centers for Disease Control and Prevention: United States surveillance data: 1997–1998 through 2009–2010 seasons http://www.cdc.gov/flu/weekly/ussurvdata.htm (2010). Accessed 18 June 2012
12. Condon, B.J., Sinha, T.: Who is that masked person: the use of face masks on Mexico City public transportation during the influenza a (H1N1) outbreak. Health Policy (2009)
13. Del Valle, S.Y., Tellier, R., Settles, G.S., et al.: Can we reduce the spread of influenza in schools with face masks? Am. J. Infect. Control **2010**, 1–2 (2010)
14. Dimitrov, N.B., Goll, S., Hupert, N., et al.: Optimizing tactics for the use of the U. S. antiviral strategic national stockpile for pandemic influenza. PLoS One **6**(1), e16094 (2011)

15. Gani, R., Highes, H., Fleming, D. et al.: Potential impact of antiviral drug use during influenza pandemic. Emerg. Infect. Dis. **11**(9) (2005)
16. Greene, V.W., Vesley, D.: Method for evaluating effectiveness of surgical masks. J. Bacteriol. **83**, 663–667 (1962)
17. Hamamura, T., Park, J.H.: Regional differences in pathogen prevalence and defensive reactions to the "Swine Flu" outbreak among East Asians and Westerners. Evol. Psychol. **8**(3), 506–515 (2010)
18. Hilleman, M.R.: Realities and enigmas of human viral influenza pathogenesis, epidemiology and control. Vaccine **20**(25–26), 3068–3087 (2002)
19. Holdren, J.P., Lander, E., Varmus, H.: Report to the president on U. S. preparations for 2009–H1N1 influenza. http://www.whitehouse.gov/assets/documents/PCAST_H1N1_Report.pdf (2009). Accessed 27 April 2012
20. Johnson, D.F., Druce, J.D., Grayson, M.L.: A quantitative assessment of the efficacy of surgical and N95 masks to filter influenza virus in patients with acute influenza infection. Clin. Infect. Dis. **2009**(49), 275–277 (2009)
21. Lee, S.A., Grinshpun, S.A., Reponen, T.: Efficiency of N95 filtering facepiece respirators and surgical masks against airborne particles of viral size range: tests with human subjects. AIHAce 2005 (2005)
22. Lee, S.A., Grinshpun, S.A., Reponen, T.: Respiratory performance offered by N95 respirators and surgical masks: human subject evaluation with NaCl aerosol representing bacterial and viral particle size range. Ann. Occup. Hyg. **52**(3), 177–185 (2008)
23. Lim, E.C., Seet, R.C., Lee, K.H., Wilder-Smith, E.P., Chuah, B.Y., Ong, B.K.: Headaches and the N95 face-mask amongst healthcare providers. Acta. Neurol. Scand. **2006**(113), 199–202 (2006)
24. Loeb, M., Dafoe, N., Mahony, J. et al.: Surgical mask vs N95 respirator for preventing influenza among health care workers. JAMA **302**(17), 1865–1871 (2009)
25. Longini, I.M., Halloran, M.E., Nizam, A., Yang, Y.: Containing pandemic influenza with antiviral agents. Am. J. Epidemiol. **2004**(159), 623–633 (2004)
26. Michaels, J.: Commercial buildings energy consumption survey. http://www.eia.doe.gov/emeu/cbecs/cbecs2003/detailed_tables_2003/detailed_tables_2003.html (2003). Accessed 12 June 2012
27. Mniszewski, S.M., Del Valle, S.Y., Stroud, P.D., et al.: Pandemic simulation of antivirals + school closures: buying time until strain-specific vaccine is available. Comput. Math. Organ. Theor. **2008**(14), 209–221 (2008)
28. Mniszewski, S.M., Del Valle, S.Y., Stroud, P.D., et al.: EpiSimS simulation of a multicomponent strategy for pandemic influenza. In: Proceedings of SpringSim, 2008
29. National Archives and Records Administration: The deadly virus: the influenza epidemic of 1918. http://www.archives.gov/exhibits/influenza-epidemic/index.html (2012). Accessed 26 April 2012
30. New York Times Health: Worry? Relax? Buy face mask? Answers on flu (2009). http://www.nytimes.com/2009/05/05/health/05well.html (2009). Accessed 26 April 2012
31. Oberg, T., Brosseau, L.M.: Surgical mask filter and fit performance. AJIC **36**(4), 276–282 (2008)
32. Osterholm, M.T.: Preparing for the next pandemic. N. Engl. J. Med. **352**(18), 1839–1842 (2005)
33. Pang, X., Zhu, Z., Xu, F., Guo, J., Gong, X., Liu, D., Liu, Z., Chin, D.P., Feikin, D.R.: Evaluation of control measures implemented in the severe acute respiratory syndrome outbreak in Beijing. J. Amer. Math. Assoc. 2003;290:3215 (2003).
34. Schoenbaum, S.C.: The impact of Pandemic Influenza, with Special Reference to 1918. International Congress Series **2001**(1219), 43–51 (2001)
35. Signorini, A., Polgreen, P.M., Segre, A.M.: The Use of twitter to track levels of disease activity and public concern in the U.S. during the influenza A H1N1 Pandemic. PLoS ONE **6**(5), e19467 (2011)
36. Stroud, P.D., Del Valle. S.Y., Mniszewski, S.M. et al.: EpiSimS pandemic influenza sensitivity analysis, part of the national infrastructure impacts of pandemic influenza phase 2 report. Los Alamos National Laboratory Unlimited Release LA-UR-07-1989 (2007)

37. Stroud, P., Del Valle, S., Sydoriak, S. et al.: Spatial dynamics of pandemic influenza in a massive artificial society. JASSS 10;4 9 http://jasss.soc.surrey.ac.uk/10/4/9.html (2007). Accessed 27 April 2012

38. Syed, Q., Sopwith, W., Regan, M., Bellis, M.A.: Behind the mask. Journey through an epidemic: some observations of contrasting public health response to SARS. J. Epidemiol. Community Health **2003**(57), 855–856 (2003)

39. U.S. Bureau of the Census: Historical U.S. population growth by year 1900–1998. Current population reports, Series P-25, Nos. 311, 917, 1095 http://www.npg.org/facts/us_historical_pops.htm (1999). Accessed 27 April 2012

40. U.S. Food and Drug Administration: Masks and N95 respirators. http://www.fda.gov/MedicalDevices/ProductsandMedicalProcedures/GeneralHospitalDevicesandSupplies/PersonalProtectiveEquipment/ucm055977.html (2010). Accessed 11 May 2012

41. U.S. News Staff: U.S. population, 2009: 305 million and counting. http://www.usnews.com/opinion/articles/2008/12/31/us-population-2009-305-million-and-counting   (2012). Accessed 27 April 2012

42. U.S. Department of Health and Human Services: The great pandemic: the United States in 1918–1919. http://www.flu.gov/pandemic/history/1918/the_pandemic/index.html. Accessed 27 April 2012

43. U.S. Department of Homeland Security: Flu pandemic morbidity/mortality. http://www.globalsecurity.org/security/ops/hsc-scen-3_flu-pandemic-deaths.htm (2011). Accessed 27 April 2012

44. U.S. Department of Transportation (DOT): Bureau of transportation statistics NHTS 2001 highlights report BTS03-05. (2003)

45. Yee, D., Bradford, J.: Employment density study. Canadian METRO Council Technical Report (1999)

46. Van Den Driessche, P., Watmough, J.: Reproduction numbers and sub-threshold endemic equilibria for compartmental models of disease transmission. Math. Biosci. **180**, 29–48 (2002)

Chapter 9
e-Epidemic Models on the Attack and Defense of Malicious Objects in Networks

Bimal Kumar Mishra and Kaushik Haldar

Abstract The efforts to develop a concrete analytical theory, which would enable us to develop a proper understanding, and then utilize the analysis in developing optimized control mechanisms to deal with the continuously increasing problem of malicious attacks in computer and other technological networks. The applicability of e-epidemic models is expected to provide a newer paradigm to the overall efforts being made in this direction. The major challenges in applying epidemic modeling to the analysis of cyber attacks have been highlighted, which are expected to provide a basic framework that can be utilized while developing effective models. The domain is expected to provide enormous challenges because of the large-scale complexities involved, but a proper design and analysis of the problem is expected to lead to comprehensive models that can possibly help in developing intelligent defense mechanisms that make use of the basic analytical framework provided by the models. The basic methodology of the applicability of e-epidemic models and their use in a qualitative and quantitative analysis of the problem has also been outlined.

1 Introduction

A major frontier that is often used to measure the success of the modern day society is the development that it has achieved in the field of communication and information technology. The rapid stride of progress that is seen in today's society cannot be imagined to survive without the backbone provided by the modern means of communication and the networks of communicating devices, that now span across the

B. K. Mishra (✉) · K. Haldar
Department of Applied Mathematics, Birla Institute of Technology,
Mesra , Ranchi 835215, India
e-mail: drbimalmishra@gmail.com

K. Haldar
e-mail: haldarkaushik@gmail.com

V. Dabbaghian and V. K. Mago (eds.), *Theories and Simulations of Complex Social Systems*, Intelligent Systems Reference Library 52,
DOI: 10.1007/978-3-642-39149-1_9, © Springer-Verlag Berlin Heidelberg 2014

globe. The growth in this sector is often argued to be comparable or even superior to that of the industrial revolution of the eighteenth century. The field of societal criminology saw huge development parallel to the large increase in the number and variations of social crime during and after the period of the industrial revolution. The field of technology related crime including cyber crime has also witnessed unparalleled growth, over the last three decades, as far as the nature, number and complexity are concerned. The worsening scenario has made it mandatory to look for efficient means that would help in developing a proper understanding of the problem, enabling us to analyze, control and frame concrete policy decisions. The continuing efforts have led to newer disciplines of research like cyber criminology and computational criminology which have developed almost simultaneously, even though they differ widely in their methodology and techniques employed. The goal of both is, however, same as both these fields are involved in ultimately developing an absolute understanding of computer related crime and then finding means to ameliorate the problem as a whole. The basic aim of this chapter is to explore the possibility of applying the principles of mathematical modeling, and in particular epidemiological modeling, towards understanding and developing a possible analytical theory of malicious attacks in technological networks. The various challenges and ramifications of modifying and effectively adapting the basic principles of biological epidemiology towards developing models for e-epidemiology will be discussed in Sect. 2.2, where not only the present scenario but the possible future directions of research in this field have been outlined. In Sects. 3 and 4, the basic methodologies applied in this domain are highlighted using a specific model. Firstly, we begin to understand the genesis of technological crime by tracing its roots to the increasing societal complexity and heterogeneity in Sect. 1.1 and then move on to the more analytical aspects of the solution in subsequent sections of the chapter.

1.1 Complex Society as the Genesis of All Crime

A popular belief is that we can never put an end to technological crime because there is often a huge lack of interest towards understanding the actual reasons and motivations behind such crime, among the actual practitioners who have to deal with security of networks at the ground level. There has however been a recent increase in the efforts to understand and relate such criminal activity to its actual roots for developing effective means that are beneficial towards improving the quality of life of the whole society,which is the actual aim of all technological improvements. In this section, we explore the direct transition that exists between the growth of societal crime, leading ultimately to technological crime, and the transition in society from uniformity towards heterogeneity. Figure 1 depicts this scenario schematically.

The initial growth of a society from a folk-based simple uniform nature, to a village society is accompanied by formation of communities. The need for survival is addressed by agriculture related occupational trends, which replace the simple nomadic, hunter-gatherer culture of the folk society. This transition leads to a cor-

Fig. 1 Direct correspondence between societal complexity and criminal activities

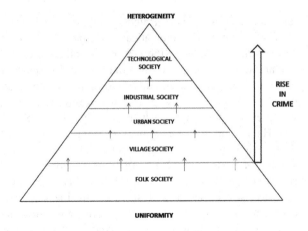

responding rise in petty criminal activities, which may include theft of cattle, crop, utensils, or other such minor crimes. The next transition from a village society to an urban society results from increasing needs and a corresponding availability of better resources and service related occupational trends. Such larger societies usually have bigger community settlements, often characterized by a similarity in occupational background, but overall cohesiveness among members of the society is often reduced. Two important characteristics, one of economic disparity among sections of the society, and the other of availability of leisure time owing to unemployment and leading to the inception of deviant behavior are often witnessed. These factors act as the root causes for an increase in the number and nature of criminal activities, which still are mostly unorganized. A further transition from an urban society to an industrial society makes the living standard of people better, but there is a corresponding rise in crime, and in particular the sudden economic changes often results in increase of organized crime. The recent formation of knowledge based technological societies has given us a quality of life which was unthinkable previously, and has led to innovations which have impacted not just the core areas of science, engineering and technology, but have allowed the common man to access and utilize the benefits of such innovations in almost every walk of life. However, the growth in complexity of crime continues to increase, reaching such levels which are not just difficult to control but even difficult to analyze.

1.2 Role of Mathematical Modeling in Study of Diseases and Crime

The control of criminal activities and that of diseases are often considered to be comparable, especially in context of their underlying basic nature, functionality and the microcosmic efforts and role of the ground level practitioners. The macrocosmic level approach towards their understanding and analysis however, has traditionally involved the use of mathematical modeling, which has proved to be considerably successful, especially in the control of epidemic diseases [1]. Their role in particular cannot be denied in the eradication of many diseases that were previously responsible for large scale human mortality. The application of mathematical methods for the study of diseases is believed to have been started by Daniel Bernoulli in the eighteenth century, when he analyzed the role of variolation against small pox. Hamer and Ross are credited for being the first to apply concrete mathematical theory when they worked on the relation between the incidence of malaria and the number of mosquitoes. The availability of modern computational techniques have added newer dimensions to the applicability of mathematical models by making them useful even in the analysis of criminal activities and also for forecasting the occurrence of crime, recognizing patterns in criminal activities and also for developing intelligent knowledge-based expert systems for controlling crimes of varying natures. The technological crimes, and in particular the malicious attacks, provide a newer challenge with greater complexities to the modern researcher.

1.3 Cyber Attacks and Mathematical Modeling

Cyber attacks provide one of the most challenging domains to cyber criminologists, one of the most interesting and practical application domains to mathematical modelers, and one of the most critical areas, on which the future of modern day technological success depends, to the computer scientist. Such attacks have been used for a variety of purposes. Macroscopically, such attacks may range across reasons like economic benefits with a purpose of fraud and forgery; societal supremacy with instances like protest-based attacks, attacks to degrade and humiliate powerful organizations, countries or even world forums; cultural and religion based reasons; terrorism-based reasons; international defence based reasons; and also to trivial curiosity and experimental reasons. Microscopically, such attacks have been used for purposes including interception and modification of data, theft and misuse of valuable information, network interference and sabotage, unauthorized access, virus and other malware dissemination, computer aided forgery and fraud. The purpose may be different, but the ever increasing numbers of such attacks on the cyber world have continued for decades to result in large scale social and economical implications. They have resulted in huge destruction of valuable information, loss of substantial amount of service and numerous hours of additional service for disinfection and recuperation of attacked systems from malicious objects. The large scale destruction caused by the attack of malicious objects in computer and other technological networks, have constantly challenged the modern day scientists to come up with innovative ways of analyzing and providing a substantial solution to the problem. The fact that the mode of attack of cyber infectives, their functionality and also their spreading characteristics are very similar to their biological counterparts, which has often led the scientific community to work with the belief that if proper mathematical models for their functioning can be developed, then an independent quantitative analysis and a qualitative comparison with their biological counterparts can help us to develop means to minimize their impact. The ultimate aim would be to move towards an analytical theory of cyber security, which on the one hand is able to meet the rigid requirements of a concrete analytical framework, and on the other hand is able to meet the practical challenging requirements needed to strengthen the ground level practitioners of the domain.

2 Development of Different e-Epidemic Models

In this section, we explore the possibilities provided by the field of epidemiology in developing methods for analyzing malicious attacks in technological networks, and then using the results obtained for developing means and methods to reduce the possible impact of such an attack. The field of biological epidemiology provides a number of useful insights into the domain, and hence we start of by looking at the interface between biological epidemiology and e-epidemiology, studying this transition, and then we briefly underline the various challenges involved, both structurally and functionally, in the process of developing e-epidemiological models.

2.1 From Biological Epidemiology to e-Epidemiology

Several models based on Epidemiology have often been applied to understand the transmission of malicious attacks in various kinds of network. The very nature of basic epidemiological modeling makes it viable as one of the only tools that can help us to suitably model and understand the system created by attacking malicious objects. It provides a dynamic modeling process where the whole population is divided into several compartments, based on their epidemiological status, and then differential or difference equations are used to specify the movements or transitions between compartments through the transition processes of infection, recovering, migration, etc. These mathematical models have traditionally been used as important tools for analyzing the transmission and control of infectious diseases. The process of model formulation has been found to be useful in clarifying assumptions, variables and parameters, while the models themselves provided conceptual results like basic reproduction numbers, contact rates, replacement numbers, etc. The mathematical models and their computer simulations have acted as useful experimental tools in building and testing theories, assessing quantitative conjectures, answering specific questions, determining sensitivities in change of parameter values and also in estimating the values of key parameters from collected data. These models have been extensively used in comparing, planning, implementing, evaluating and optimizing different programs for detection, prevention, therapy and control of epidemics. Structurally and functionally, most epidemic models can be traced to the classical *Susceptible-Infectious-Recovered* (SIR) model proposed by Kermack and McKendrick [2] but the first complete application of mathematical models to computer virus propagation using epidemic models was based on directed graphs for virus transmission [3]. This marked the beginning of the application of epidemiological modeling to computer networks, with a number of models being developed to deal with various aspects of such epidemics, mostly considering the similarities between such infective in the biological world and in the cyber world. The major strength and support behind the applicability of such models comes not just from the similarities between the infecting objects, but also from certain other relevant factors. One major factor is that in most sciences, it is possible to conduct experiments to obtain information and test hypotheses, but experiments with the spread of viruses, worms or other malware in computer networks are often impossible, unethical and expensive, and also the data publicly available is often incomplete because of under-reporting. The lack of reliable data makes accurate estimation of parameters very difficult and as such it may only be possible to estimate a range of values of some of the parameters. Thus the mathematical models and computer simulations provide the necessary tools to perform the possible theoretical experiments to abridge the gaps caused by the non-availability of practical experiments.

2.2 Structural and Functional Implications and Major Challenges in e-Epidemic Modeling

The applicability of epidemic modeling has opened a newer dimension in the analysis of the attack and defense of malicious objects in computer and other technological networks. To make the use of such modeling useful and advantageous to the overall goal of an improved theory of computer security, it is vital to understand the varying implications of e-epidemic modeling and also to identify the major challenges of the domain. In this section, the chapter deals with these aspects of e-epidemic modeling.

2.2.1 Characterization of Nodes Based on the Status of Infection

The basic pre-criterion for applying an epidemic model based analysis, is to characterize and classify the nodes into the infected and non-infected categories, depending on the presence or absence of infection. The basic models of epidemiology [2] provide the foundational structure for such models. Two possible approaches may arise depending on whether there is a distinction made between the non-infected nodes that are yet to be infected and those which have recovered from the infection. In both the methods, the initially non-infected nodes are considered in a *susceptible* class, while the infected nodes are taken in an *infected* class. In the model with no difference being made between the non-infected and recovered nodes, the infected nodes after recovery move back into the susceptible class, while in the other approach, a sep-

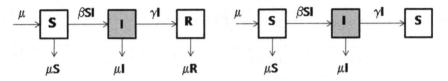

Fig. 2 Schematic representation of **a** Susceptible - Infected-Recovered model, and **b** Susceptible-Infected-Susceptible model

Table 1 Nomenclature

Symbol	Description
$S(t)$	The susceptible population size
$E(t)$	The exposed population size
$I(t)$	The infectious population size
$Q(t)$	The quarantined population size
$R(t)$	The recovered population size
$N(t)$	The total population size
β	The per infectivity contact rate
γ	The rate of recovery of infectious nodes
μ	The natural death rate and birth rate of nodes
ε	The rate at which exposed nodes become infectious
ξ	The rate at which recovered nodes lose their temporary immunity
δ	Death rate due to infection
ν	Rate at which quarantined nodes recover
ψ	Rate of vaccination of susceptible nodes

arate *recovered* class is considered. The nomenclature of the models depending on the dynamics of the infection becomes Susceptible-Infected-Recovered (SIR) model and Susceptible-Infected-Susceptible (SIS) model, based on the presence or absence of the recovered class.

The functionality of these models follows directly from biological epidemiology, and their efficacy in the domain of computer networks, is mainly to provide a black-box structure that serves as the basic foundation on which almost all of e-epidemiological modeling is based. The shaded infected class in 2 is mainly formatted to provide the possibility of inclusion of the various facets that need to be considered while applying epidemiological models to technological attacks.

The functional dynamics of the two models can be represented using the following sets of differential equations.

Model 1(Susceptible-Infected-Recovered (SIR)):

$$\frac{dS}{dt} = \mu - \beta SI - \mu S$$

$$\frac{dI}{dt} = \beta SI - (\mu + \gamma)I \tag{1}$$

$$\frac{dR}{dt} = \gamma I - \mu R$$

Model 2 (Susceptible-Infected-Susceptible (SIS)):

$$\frac{dS}{dt} = \mu - \beta SI - \mu S + \gamma I$$

$$\frac{dI}{dt} = \beta SI - (\mu + \gamma)I \tag{2}$$

2.2.2 Characterization of Nodes Based on the Status of Infectivity

A major structural variation also arises when a differentiation is made between nodes that are infected but are not yet able to spread the infection to other nodes, and those infected nodes that already have the ability to infect other nodes [4]. This difference structurally leads to the formation of two infected classes, by partitioning of the previously single infected class, where the two classes are named as the *exposed* class and the *infectious* class. Functionally, this allows the inclusion of a latent period of infection, that accounts for the time gap between a node getting infected to the point where it starts to spread the infection to other nodes. This period is generally not very high in case of an attack in a technological network, but it often acts as an important feature of such networks whose inclusion is found to be useful in understanding the functionality of such attacks.However, in many malicious attacks, a delay is purposefully built into the functionality. For instance, an arriving worm may need time to transfer its malicious payload to a host node. Moreover, the small latency period may be negligible in terms of the total infection span, like those of many fast spreading worms like Code Red or Slammer worm. Figure 3 shows how the model structurally appears, when the infected class of model 2 is partitioned to include the latently infected *exposed* class.

The model is referred to as the Susceptible-Exposed-Infectious-Recovered (SEIR) model and its dynamics can be represented by the following system of equations.

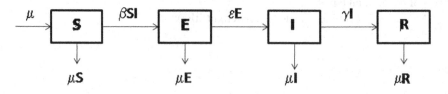

Fig. 3 Schematic representation of Susceptible-Exposed-Infectious-Recovered model

Model 3(Susceptible-Exposed-Infectious-Recovered (SEIR)):

$$\frac{dS}{dt} = \mu - \beta SI - \mu S$$
$$\frac{dE}{dt} = \beta SI - (\mu + \varepsilon)E$$
$$\frac{dI}{dt} = \varepsilon E - (\mu + \gamma)I \tag{3}$$
$$\frac{dR}{dt} = \gamma I - \mu R$$

2.2.3 Characterization of Nodes Based on the Nature of Immunity

The structure of the models can also be further modified, considering whether the nodes on recovery acquire permanent immunity to the infection or the immunity is only for a specific period of time after which it is lost [4]. The structural implication of the loss of immunity is in the form of a chain of classes, where the nodes enter back into the susceptible class, once they lose the acquired immunity. From the functional point of view, this feature turns out to be useful in modeling the attack of particular strains of attacking agents, where upon using an updated antivirus which can identify the signature of the particular agent, the immunity can be safely assumed to be permanent, whereas the contrary assumption holds when a whole class of evolving attacking agents or a sub-class of sufficient size is taken into consideration. In Fig. 4, model 4 is extended to include the case of temporary immunity, while model 4 itself considers the assumption of permanent immunity.

The model is named as the Susceptible-Exposed-Infectious-Recovered-Susceptible (SEIRS) model, which includes a transition of nodes from the recovered class back to the susceptible class. The dynamics of the model can be represented by the following system of equations.

Model 4 (Susceptible-Exposed-Infectious-Recovered-Susceptible (SEIRS)):

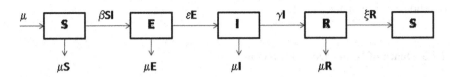

Fig. 4 Schematic representation of Susceptible-Exposed-Infectious-Recovered-Susceptible (SEIRS) model which assumes temporary immunity to infection

$$\frac{dS}{dt} = \mu - \beta SI - \mu S + \xi R$$

$$\frac{dE}{dt} = \beta SI - (\mu + \varepsilon)E$$

$$\frac{dI}{dt} = \varepsilon E - (\mu + \gamma)I \tag{4}$$

$$\frac{dR}{dt} = \gamma I - (\mu + \xi)R$$

2.2.4 Natural Death of Nodes

The removal of non-functional nodes from the network based on reasons not related to the attack of malicious objects is often referred to as *natural death* or *natural mortality*, borrowing terminology from biological epidemiology. In epidemiological modeling of technological attacks, there are again functionally two different aspects to be considered when deciding on the nature of the natural death rate. The rate can be sometimes assumed to be a function of time to account for the fact that the nodes in a computer network have a life-cycle like any other machine, where the node may be assumed to have zero death rate up to its expected lifetime and then it is assumed to be exponentially increasing. Assuming 'a' to be the expected lifetime, the death rate may be defined as follows

$$\mu(t) = \begin{cases} 0; & t < a \\ ce^{dt}; & t \geq a \end{cases} \tag{5}$$

where c and d are positive constants. The preferred choice for the natural death rate is however to assume it to be a small constant because in general the modeled time span for an attack and its subsiding is very small, when compared to the average life time of the nodes. Therefore for a short time modeling process, the constant choice of natural death rate suffices for most of e-epidemic modeling, but if there is a need to model a time range which is larger than the expected life span of the nodes, then the choice can be made as in Eq. 5.

2.2.5 Death of Nodes Due to Infection

The loss of functionality of nodes due to the attack of malicious objects is often referred to as the *infection-induced death* or *infection-induced mortality* or *crashing*. In most active attacks in computer networks, the primary aim is to damage the nodes to an extent that they become non-functional, whereas in most passive attacks, the aim is only to find out vulnerable nodes which can be later compromised to launch active internal attacks. The role of infection-induced death rate is therefore limited in case of modeling of passive attacks, whereas it is vital to consider its impact in case

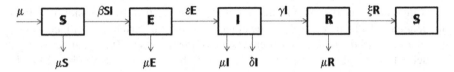

Fig. 5 Schematic representation of Susceptible-Exposed-Infectious-Recovered-Susceptible (SEIRS) model with infection-induced death in Infectious (I) class

of active attacks, where there is substantial crashing of nodes due to the attack. The corresponding death rate is assumed in one or more of the infected classes, where the rates may be assumed to be same for computational simplicity, or different, in which case the death rate in the infectious (I) class is assumed to be more than the others to show that it is the primary class as far as deaths are concerned, among all the infected classes. The possibility of a variable death rate is not considered in this case because of the instantaneous nature of the deaths in an overwhelming majority of the cases. In Fig.5, an extended model obtained by considering infection-induced death in the infectious class of model 5, is shown

The corresponding model is same as model 5, except for the equation representing the dynamics of the infectious class, which gets modified to

$$\frac{dI}{dt} = \varepsilon E - (\mu + \gamma + \delta)I \tag{6}$$

2.2.6 Vertical Transmission of Infection

The concept of vertical transmission has been studied in a number of biological epidemic models to account for the fact that a fraction of the off-springs of infected hosts will already be infected at birth. The idea proves useful even in e-epidemiological modeling, where new terminals added to an infected node or all nodes that are connected to an infected hub, have a very high probability of getting directly infected. Figure 6 shows an extension of the SEIRS model, with addition of infected nodes to the exposed (E) class, which includes both the fraction of nodes added to the exposed nodes ($p\mu E$) and also the infectious nodes ($q\mu I$).

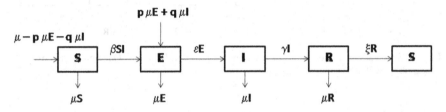

Fig. 6 Schematic representation of Susceptible-Exposed-Infectious-Recovered-Susceptible (SEIRS) model with vertical transmission

A qualitative analysis shows that the addition of vertical transmission into the model functionally increases the value of the threshold parameter, thereby increasing the chance of formation of the endemic state of the infection. The dynamics of the model get modified with respect to the rates of change of the susceptible (S) and exposed (E) classes, whose equations become

$$\frac{dS}{dt} = \mu - p\mu E - q\mu I - \beta SI - \mu S + \xi R$$
$$\frac{dE}{dt} = \beta SI + p\mu E + q\mu I - (\mu + \varepsilon)E \qquad (7)$$

2.2.7 Application of Quarantining of Infected Nodes

The idea of forcefully removing a section of the infected nodes from the network to reduce the rate of spread of further infection to non-infected nodes is often referred to as *quarantining*, and it serves the dual purpose of preventing the infected nodes from coming into active contact with other nodes and also allows the model to account for the time period needed to disinfect the already infected nodes. The inclusion of a *quarantined* class functionally reduces the spread of infection and acts as a good abstraction of a practical process that is often employed in most computer networks [5]. Structurally, the model gets modified by including the extra quarantined (Q) class before the nodes get recovered, and schematically it would appear as shown in Fig. 7.

The model once again is similar in qualitative aspects with the previous form, where quarantining has not been considered, but in effect the features of the mechanism are seen to show an impact as far as the level of infection and its spread are concerned. The system of equations representing the model with quarantining included is as follows

Model 5 (Susceptible-Exposed-Infectious-Quarantined-Recovered-Susceptible (SEIQRS)):

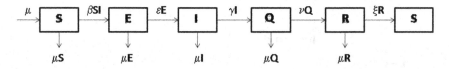

Fig. 7 Schematic representation of Susceptible-Exposed-Infectious-Quarantined-Recovered-Susceptible (SEIQRS) model

$$\frac{dS}{dt} = \mu - \beta SI - \mu S + \xi R$$

$$\frac{dE}{dt} = \beta SI - (\mu + \varepsilon)E$$

$$\frac{dI}{dt} = \varepsilon E - (\mu + \gamma)I \tag{8}$$

$$\frac{dQ}{dt} = \gamma I - (\mu + v)Q$$

$$\frac{dR}{dt} = vQ - (\mu + \xi)R$$

2.2.8 Strength of Defense Mechanism Involved

A major practical feature on which the success of a network security mechanism depends is the strength of the defense mechanism involved, which includes both the strength of the anti-malicious and intrusion detection software and also the strength of the manual technicians working, which specially plays a considerable role when the modeled network is a large enterprise network and the resources spent during the disinfection process cannot be ignored. The strength of the defense mechanism is often represented by the rate at which the nodes make a transition from the infectious (I) class into the recovered (R) class, when direct transition is considered. On the other hand, if quarantining is taken into consideration then it is represented by a combination of the rate of transition from the infectious (I) class to the quarantined (Q) class along with the rate of transition from the quarantined (Q) class to the recovered (R) class. These rates can in particular be useful to apply the model for a quantitative analysis of the model, where the costs incurred in terms of the resources used or wasted can be estimated.

2.2.9 Impact of Control Techniques

To reduce the effective rate of spread of infection in a network of a malicious attack, various kinds of control mechanisms can be devised. An e-epidemic model may need to represent a number of such features or actions in the physical system, like presence or absence of antivirus, its relative strength, availability of virus signatures, effectiveness of antivirus to deal with a certain class of attacking agents, relative degradation of the antivirus mechanism when considering evolving attacking agents, among a number of other possibilities. One of the most commonly used control mechanisms is often referred to as vaccination, which may be used to represent a number of the stated scenario, under different kinds of assumptions. Figure 8 represents two varying kinds of situations in which the process of vaccination can be applied in e-epidemic models. In the first case (Fig. 8a), the underlying model assumes permanent immunity and so the vaccinated nodes are assumed to move directly to the recovered (R) class. In the other situation (Fig. 8b), immunity is temporary and so

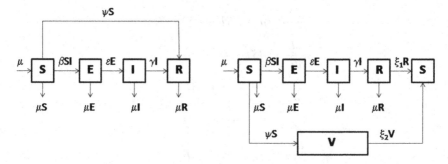

Fig. 8 Schematic representation of **a** SEIR (permanent immunity) and **b** SEIRS (temporary immunity) models with vaccination

not just the recovered nodes lose their immunity but also the directly vaccinated nodes lose the acquired immunity and again become susceptible to the attack. The permanent immunity concept has limited applicability, specially when modeling a particular strain of malicious object. However, in most cases the immunity is lost as soon as the attacking malware evolves, or a new type of malicious object comes into the scenario.

The term ψS representing the susceptible nodes which become directly vaccinated, gets reduced from the equation for the susceptible (S) class and get added into the recovered (R) class in the SEIR model in the case of permanent immunity. In the model with temporary immunity, a new vaccinated (V) class is introduced whose rate of dynamics is given by the equation

$$\frac{dV}{dt} = \psi S - \xi_2 V \tag{9}$$

while the modified equation for the susceptible (S) class is

$$\frac{dS}{dt} = \mu - \beta SI - \mu S + \xi_1 R + \xi_2 V \tag{10}$$

Here ξ_1 and ξ_2 represent the rates at which the recovered nodes and the vaccinated nodes lose their temporary immunity to the infection.

2.2.10 Impact of Non-linearity in Incidence

In all the models discussed so far, the rate of formation of new infections or the incidence rate is assumed to be bilinear, depending linearly on the fraction of infectious nodes and the fraction of susceptible nodes, which has a direct consequence of the underlying simplifying assumption of homogeneous mixing. For e-epidemiological models, this type of incidence may need to be modified depending on the topology of the type of network being modeled. The bilinear incidence may hold in a number of simplified situations or in case of networks which have a mesh or a nearly mesh topology, where the principle of homogeneous mixing stands justified. The applicability however is limited to specific scenarios like ad hoc and sensor networks but in most practical cases, including the Internet, the contacts of any node with other nodes are highly limited by the topology, which makes homogeneity an overly simplified assumption. The Internet may appear to a user to be a mesh network but a graphical representation of it is always a highly sparsely connected structure. A node may still be used for scanning other nodes to find vulnerable hosts on an IPv4 network but even this advantage is expected to be diluted with the change over to 128 bit addresses in IPv6 networks. A logical extension is to assume a non-linear incidence where the rate depends non-linearly, given by positive parameters p and q, on both the infectious and susceptible classes of nodes. The SEIQRS model, discussed earlier can be modified by including bilinear incidence to have the following system of equations:

Model 6 (Susceptible-Exposed-Infectious-Quarantined-Recovered-Susceptible (SEIQRS) model with non-linear incidence):

$$\frac{dS}{dt} = \mu - \beta S^p I^q - \mu S + \xi R$$
$$\frac{dE}{dt} = \beta S^p I^q - (\mu + \varepsilon) E$$
$$\frac{dI}{dt} = \varepsilon E - (\mu + \gamma) I \tag{11}$$
$$\frac{dQ}{dt} = \gamma I - (\mu + v) Q$$
$$\frac{dR}{dt} = v Q - (\mu + \xi) R$$

2.2.11 Characterization of Different Behavior of Epidemic Sub-classes

An e-epidemic model may need to show structural and functional differentiality with respect to the epidemic classes, leading to the formation of sub-classes, to represent various kinds of physical situations like difference in kinds of attacking agents leading to differential—susceptibility models, difference in exhibited symptoms leading to differentially—exposed models, difference in infectiousness leading to differentially—infectious models and difference in methods of disinfection leading to differentially—quarantined models [6–8]. These models can provide useful insight towards understanding technological attacks, and towards better practical modeling of the scenario involved. Such models can also be effectively employed for effective applications like vulnerability analysis of a network for different kinds of infections, symptom based analysis and detection for different kinds of behavior during latent infection phase where nodes are still to become infectious and so can be checked in time, for performance analysis of the defense mechanism for different sets of defense mechanisms or teams of manual technicians, or a cost based analysis to

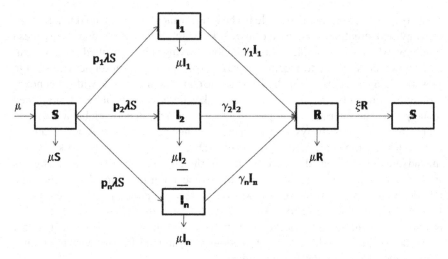

Fig. 9 Schematic representation of differentially infectious SIRS model

optimize the way that a large enterprise network may schedule the use of its resources to meet its security requirements. Figure 9 shows the structural representation of a differentially-infectious SIRS model.

The model may be represented by the following system of equations

Model 7 (Differentially Infectious SIRS model):

$$\frac{dS}{dt} = \mu - \lambda S - \mu S + \xi R$$

$$\frac{dI_i}{dt} = p_i \lambda S - (\mu + \gamma_i)I_i; \quad i = 1, 2, ..., n \tag{12}$$

$$\frac{dR}{dt} = \sum_{i=1}^{n} \gamma_i I_i - (\mu + \xi)R$$

where λ represents the incidence and p_i represents the probability that a susceptible node gets infected into the ith infectious class. The nodes may show symptoms of more than one group or may be infected by more than one type of attacking agent, in most practical situations. However, the utility of such models would depend on a suitable choice of method to decide the dominant sub-class.

This section highlighted the various facets of e-epidemic modeling, with particular reference to the context of malicious attacks in networks. However, there needs to be a careful consideration of the context in which the attack is being modeled. For instance, when modeling attacks of specific virus strains, the concept of permanent recovery is considered, unless the time frame is large enough to include mutated variations of the attacking agent. The models with temporary recovery are expected to be applicable when modeling attacks by general classes of attacking agents, or if there is a provision for a strengthened attack from the same agent. Moreover, models including differential behaviour allow the consideration of multiple classes of attacking agents together. The choice therefore needs to be careful enough to enable the model to be as practically viable and relevant as possible, based on the scenario being modeled. Moreover, considering the growing complexity of technological networks, the aim for researchers in the domain becomes developing models that are comprehensive enough to include features of such networks as well, and if required to develop specific models for such networks, with varying assumptions and possibly varying inferences that would be helpful in dealing with such attacks. The challenges are also being compounded by the ever-increasing complexity of the nature and number of attacks being faced by the networks worldwide, which makes it essential to plan and develop more comprehensively applicable models, with possible provisions for feature selection to address specific requirements. Even though these challenges are substantial, the possibilities of arriving at a concrete analytical theory which could possibly help towards effectively solving the problem of malicious attacks in technological networks. In the next section, the qualitative aspects of e-epidemiological modeling will be explored.

3 Equilibria and Their Stability

In this section, the essential qualitative features of epidemiological models will be explored briefly, which would include a threshold parameter called the *basic reproduction number* (R_0) and the conditions specified by it, that determines whether the infection persists in the network asymptotically or it eventually dies out with time. For the analysis, a generalized SEIQR (Susceptible-Exposed-Infectious-Quarantined-Recovered) model with infection induced deaths will be considered throughout this section. The infection induced deaths are considered in both the infectious (I) class and the quarantined (Q) class, with the rates δ_1 and δ_2 respectively, being different in the two classes. The structural representation of the model is as shown in Fig. 10.

The system of equations forming the model is as follows Model 8 (Susceptible-Exposed-Infectious-Quarantined-Recovered (SEIQR) with infection induced deaths):

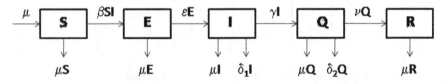

Fig. 10 Schematic representation of Susceptible-Exposed-Infectious-Quarantined-Recovered (SEIQR) model with infection-induced death in Infectious (I) and Quarantined (Q) class

$$\frac{dS}{dt} = \mu - \beta SI - \mu S$$

$$\frac{dE}{dt} = \beta SI - (\mu + \varepsilon)E$$

$$\frac{dI}{dt} = \epsilon E - (\mu + \delta_1 + \gamma)I \qquad\qquad (13)$$

$$\frac{dQ}{dt} = \gamma I - (\mu + \delta_2 + v)Q$$

$$\frac{dR}{dt} = vQ - \mu R$$

3.1 Basic Reproduction Number and Local Stability Conditions of Infection Free Equilibrium

The basic reproduction number R_0 is one of the most important tool for epidemiologists, as it gives a measure of the expected number of secondary infections produced by a node during its total infectious period, when introduced in a population of susceptible nodes [9]. The definition of R_0 immediately gives us a threshold condition, because when $R_0 > 1$, each infected node during its lifetime infects more than one other node on an average, and so the epidemic persists, whereas the condition gets reversed when $R_0 \leq 1$ and the infection dies out. The infection free equilibrium for the above system is given as $(S = 1, E = 0, I = 0, Q = 0, R = 0)$, and the most important applicability of R_0 is that the stability of this equilibrium is clearly reflected. To obtain the basic reproduction number R_0, the next generation method [10, 11] is used. This method is the most suitable method for deriving R_0 in models that include multiple classes of infected nodes. In this method, the spectral radius, or the supremum of the absolute values of the spectral elements, of the operator is obtained. The spectrum here refers to the set of all the eigen values of the operator. The matrix F representing the rate of appearance of new infections and the matrix V representing the difference between inward and outward flow of nodes into a compartment are given as

$$F = \begin{pmatrix} \beta SI \\ 0 \\ 0 \\ 0 \\ 0 \end{pmatrix} \quad \text{and} \quad V = \begin{pmatrix} (\mu + \varepsilon)E \\ -\varepsilon E + (\mu + \delta_1 + \gamma)I \\ \beta SI - \mu + \mu S \\ -\gamma I + (\mu + \delta_2 + v)Q \\ -vQ + \mu R \end{pmatrix}$$

Taking the partial derivatives with respect to the infectious classes, we get

$$F = \begin{pmatrix} 0 & \beta SI \\ 0 & 0 \end{pmatrix} \quad \text{and} \quad V = \begin{pmatrix} \mu + \varepsilon & 0 \\ -\varepsilon & \mu + \delta_1 + \gamma \end{pmatrix}$$

The basic reproduction number is given as the spectral radius (ρ) of the next generation operator FV^{-1} and so we have $R_0 = \rho(FV^{-1})$ where the matrix FV^{-1} is given as

$$FV^{-1} = \frac{1}{(\mu + \epsilon)(\mu + \delta_1 + \gamma)}A$$

where

$$A = \begin{pmatrix} \beta\epsilon & \beta(\mu + \epsilon) \\ 0 & 0 \end{pmatrix}$$

at the infection free equilibrium. So, the value of R_0 becomes

$$R_0 = \frac{\beta\epsilon}{(\mu + \epsilon)(\mu + \delta_1 + \gamma)} \tag{14}$$

This value of the basic reproduction number acts as a threshold on which the formation or failure of the epidemic event depends as already discussed, and the stability of the infection free equilibrium also depends on the condition, which is summarized in the following subsection. From the definition of R_0, the infection free equilibrium is locally asymptotically stable if $R_0 \leq 1$, and is unstable if $R_0 > 1$. The values of basic reproduction number for some of the epidemic models which were discussed earlier in Sect. 2.2 are enlisted in Table 2.

Next, in Sects. 3.2 and 3.3 the global stability conditions for the infection free and endemic equilibrium points are obtained. These conditions broadly specify the criteria under which the persistence or failure of the epidemic depend and hence allow the determination of the basic path to be adopted to deal with an epidemic malicious attack.

3.2 Global Stability of Infection Free Equilibrium

The LaSalle's invariance principle can be used to show that the condition for the global asymptotic stability of the infection free equilibrium point in the domain $\Gamma = \{(S, E, I, Q, R) : S \geq 0, E \geq 0, I \geq 0, Q \geq 0, R \geq 0, S+E+I+Q+R \leq 1\}$ is that $R_0 \leq 1$, i.e. all solutions starting in this feasible region approach the infection free equilibrium when $R_0 \leq 1$. Here we need to select a real-valued function L with domain Γ, which has properties analogous to the potential function in classical dynamics, and its choice has to consider the transitions for the infected classes. We define such a function L as $L = \epsilon E - (\mu + \epsilon)I$.

Then the time derivative of L is given as $L' = (\mu+\epsilon)(\mu+\delta_1+\gamma)[\frac{\beta\epsilon S}{(\mu+\epsilon)(\mu+\delta_1+\gamma)} - 1]I = (\mu + \epsilon)(\mu + \delta_1 + \gamma)(R_0 S - 1)I$.

This shows that $L' \leq 0$ if $R_0 \leq 1$ and also $L' = 0$ if and only if I = 0 or $R_0 = 1$ and S = 1. Hence the largest compact invariant set in Γ where $L' = 0$ is the singleton set containing the infection free equilibrium point, and so it is globally stable in Γ. The next section deals with the endemic equilibrium and conditions for its stability.

Table 2 Basic epidemic models and their properties

Sl. No.	Model	Basic reproduction number	Fundamental properties
1	SIR	$\frac{\beta}{\gamma+\mu}$	1. Single infected class
			2. Permanent immunity
2	SIS	$\frac{\beta}{\gamma+\mu}$	1. Single infected class
			2. Temporary immunity
3	SEIR	$\frac{\beta\epsilon}{(\gamma+\mu)(\epsilon+\mu)}$	1. Presence of latently infected class
			2. Permanent immunity
4	SEIRS	$\frac{\beta\epsilon}{(\gamma+\mu)(\epsilon+\mu)}$	1. Presence of latently infected class
			2. Temporary immunity
5	SEIQR	$\frac{\beta\epsilon}{(\gamma+\mu)(\epsilon+\mu)}$	1. Presence of latently infected class
			2. Permanent immunity
			3. Presence of quarantine based disinfection phase
6	SEIQRS	$\frac{\beta\epsilon}{(\gamma+\mu)(\epsilon+\mu)}$	1. Presence of latently infected class
			2. Temporary immunity
			3. Presence of quarantine based disinfection phase
7	SEIQR	$\frac{\beta\epsilon}{(\mu+\epsilon)(\mu+\delta_1+\gamma)}$	1. Presence of latently infected class
	(with infection induced deaths)		2. Permanent immunity
			3. Presence of quarantine based disinfection phase
			4. Presence of infection induced deaths in I and Q classes

3.3 Global Stability of Endemic Equilibrium

In this last section, it is observed that the system is asymptotically converging to the infection free state when $R_0 \leq 1$. However, when $R_0 > 1$ then the infection free equilibrium loses its global stability and an endemic equilibrium exists, which guarantees that under this situation the infection persists in the network. The endemic equilibrium point for the SEIQR model with infection induced deaths satisfies the following system of equations

$$l\mu - \beta SI - \mu S = 0$$
$$\beta SI - (\mu + \epsilon)E = 0$$
$$\epsilon E - (\mu + \delta_1 + \gamma)I = 0 \qquad (15)$$
$$\gamma I - (\mu + \delta_2 + v)Q = 0$$
$$vQ - \mu R = 0$$

Solving these equations, we get the endemic equilibrium point $(S^*, E^*, I^*, Q^*, R^*)$ as

$S^* = \frac{1}{R_0}$

$E^* = \frac{\mu}{\mu+\epsilon}(1 - \frac{1}{R_0})$

$I^* = \frac{\mu(R_0-1)}{\beta}$

$Q^* = \frac{\gamma\mu(R_0-1)}{\beta(\mu+\delta_2+v)}$

$R^* = \frac{\gamma v(R_0-1)}{\beta(\mu+\delta_2+v)}$

We next adapt a geometric approach [12, 13] to show the global stability of the endemic equilibrium. According to their approach for the mapping $f : D \subset \Re^n \to \Re^n$, where D is an open set, if the differential equation $x' = f(x)$ be such that its every solution x(t) can be uniquely determined by its initial condition $x(t) = x_0$, then an equilibrium point $\bar{x} \epsilon D$ and satisfying the conditions: (1) D is simply connected, (2) There exists a compact absorbing subset K of D, (3) $\bar{x}$ is the only equilibrium point in D, is globally stable if it satisfies the additional Bendixson criteria given as

$\bar{q_2} = \lim \sup_{t\to\infty} \sup_{x_0\epsilon K} q < 0$

where

$q = \int_0^t \mu(B(x(s, x_0)))\, ds$

Also $B = A_f A^{-1} + A\frac{\partial f^{[2]}}{\partial x} A^{-1}$ and A is a matrix-valued function satisfying $\mu(A_f A^{-1} + AJ^{[2]}A^{-1}) \leq -\delta < 0$ on K. Further $J^{[2]} = \frac{\partial f^{[2]}}{\partial x}$ is the second compound Jacobian matrix [14] which for a Jacobian matrix of order 4 is given

as $J^{[2]} = \begin{pmatrix} j_{11}+j_{22} & j_{23} & j_{24} & -j_{13} & -j_{14} & 0 \\ j_{32} & j_{11}+j_{33} & j_{34} & j_{12} & 0 & -j_{14} \\ j_{42} & j_{43} & j_{11}+j_{44} & 0 & j_{12} & j_{13} \\ -j_{31} & j_{21} & 0 & j_{22}+j_{33} & j_{34} & -j_{24} \\ -j_{41} & 0 & j_{21} & j_{42} & j_{22}+j_{44} & j_{23} \\ 0 & -j_{41} & j_{31} & -j_{42} & j_{32} & j_{33}+j_{44} \end{pmatrix}$ and

μ denotes the Lozinskii measure, given as $\mu(M) = \lim_{h\to 0+} \frac{|I+hM|-1}{h}$ for an N X N matrix M.

The existence of a compact absorbing set which is absorbing in the interior of Γ, follows from the uniform persistence of the system as $\lim \inf_{t\to\infty} S(t) > c$, $\lim \inf_{t\to\infty} E(t) > c$, $\lim \inf_{t\to\infty} I(t) > c$, $\lim \inf_{t\to\infty} Q(t) > c$ and $\lim \inf_{t\to\infty} R(t) > c$ for some $c > 0$. Based on the procedure used by [13] and later by [12], the proof for the Bendixson criteria $\overline{q_2} < 0$, can be enumerated in the form of the following steps:

(1) The Jacobian matrix of the reduced system, leaving the recovered class is

$$J = \begin{pmatrix} -\mu - \beta I & 0 & -\beta S & 0 \\ \beta I & -(\mu + \varepsilon) & \beta S & 0 \\ 0 & \varepsilon & -(\mu + \delta_1 + \gamma) & 0 \\ 0 & 0 & \gamma & -(\mu + \delta_2 + \nu) \end{pmatrix}$$

(2) The second compound additive Jacobian matrix is given as

$$J^{[2]} = \begin{pmatrix} -a & \beta S & 0 & \beta S & 0 & 0 \\ \varepsilon & -b & 0 & 0 & 0 & 0 \\ 0 & \gamma & -c & 0 & 0 & -\beta S \\ 0 & \beta I & 0 & -d & 0 & 0 \\ 0 & 0 & \beta I & \gamma & -e & \beta S \\ 0 & 0 & 0 & 0 & \varepsilon & -f \end{pmatrix}$$

where
$a = (2\mu + \beta I + \varepsilon),\ b = (2\mu + \beta I + \delta_1 + \gamma),\ c = (2\mu + \beta I + \delta_2 + \nu),$
$d = (2\mu + \gamma + \delta_1 + \varepsilon),\ e = (2\mu + \varepsilon + \delta_2 + \nu),\ f = (2\mu + \delta_1 + \delta_2 + \gamma + \nu)$

(3) To obtain matrix B in the Bendixson criteria, we define a diagonal matrix A as

$$A = diag(1, \tfrac{E}{T}, \tfrac{E}{T}, \tfrac{E}{T}, \tfrac{E}{T}, \tfrac{E}{T})$$

and so if f denotes the vector field of the system, then

$$A_f A^{-1} = diag(0, (\tfrac{E}{T})_f \tfrac{I}{E}, (\tfrac{E}{T})_f \tfrac{I}{E}, (\tfrac{E}{T})_f \tfrac{I}{E}, (\tfrac{E}{T})_f \tfrac{I}{E}, (\tfrac{E}{T})_f \tfrac{I}{E})$$

Hence the matrix B is given as

$$B = A_f A^{-1} + A J^{[2]} A^{-1} = \begin{pmatrix} -a & \frac{\beta SI}{E} & 0 & \frac{\beta SI}{E} & 0 & 0 \\ \frac{E\varepsilon}{T} & (\frac{E}{T})_f \frac{I}{E} - b & 0 & 0 & 0 & 0 \\ 0 & \gamma & (\frac{E}{T})_f \frac{I}{E} - c & 0 & 0 & -\beta S \\ 0 & \beta I & 0 & (\frac{E}{T})_f \frac{I}{E} - d & 0 & 0 \\ 0 & 0 & \beta I & \gamma & (\frac{E}{T})_f \frac{I}{E} - e & \beta S \\ 0 & 0 & 0 & 0 & \varepsilon & (\frac{E}{T})_f \frac{I}{E} - f \end{pmatrix}$$

which can be written in the form of a block matrix as $B = \begin{pmatrix} B_{11} & B_{12} \\ B_{21} & B_{22} \end{pmatrix}$ where

$B_{11} = (-a)$

$B_{12} = \begin{pmatrix} \frac{\beta SI}{E} & 0 & \frac{\beta SI}{E} & 0 & 0 \end{pmatrix}$

$B_{21} = \begin{pmatrix} \frac{E\gamma}{T} \\ 0 \\ 0 \\ 0 \\ 0 \end{pmatrix}$

$$
B_{22} = \begin{pmatrix}
(\frac{E}{I})_f \frac{I}{E} - b & 0 & 0 & 0 & 0 \\
\gamma & (\frac{E}{I})_f \frac{I}{E} - c & 0 & 0 & -\beta S \\
\beta I & 0 & (\frac{E}{I})_f \frac{I}{E} - d & 0 & 0 \\
0 & \beta I & \gamma & (\frac{E}{I})_f \frac{I}{E} - e & \beta S \\
0 & 0 & 0 & \varepsilon & (\frac{E}{I})_f \frac{I}{E} - f
\end{pmatrix}
$$

(4) The Lozinskii measure of matrix B can be estimated as $\mu(B) \leq \sup\{g_1, g_2\}$
where g_1 and g_2 are defined as
$g_1 = \mu(B_{11}) + |B_{12}| = -2\mu - \beta I - \varepsilon + \frac{\beta SI}{E}$
and $g_2 = |B_{21}| + \mu_1(B_{22}) = -2\mu + \frac{I}{E}\frac{\varepsilon E}{I} + (\frac{E}{I})_f$
where, the Lozinskii measure mu_1 is with respect to the l_1 norm and the norms
of matrices B_{12} and B_{21} are the operator norm.
Now $\frac{I}{E}\frac{\varepsilon E}{I} + (\frac{E}{I})_f = \frac{I}{E}(\frac{IE' - EI'}{I^2}) = \frac{E'}{E} - \frac{I'}{I}$ and so
$\frac{E'}{E} = \frac{\beta SI}{E} - (\mu + \varepsilon)$ and $\frac{I'}{I} = \frac{\varepsilon E}{I} - (\mu + \delta_1 + \gamma)$.
Hence g_1 and g_2 reduce to $g_1 = -\mu - \beta I + \frac{E'}{E}$ and $g_2 = -\mu + \gamma + \delta_1 + \frac{E'}{E}$.
So, $\mu(B) \leq \sup\{g_1, g_2\} \leq \frac{E'}{E} - \mu + \sup\{-\beta I, \delta_1 + \gamma\}$
which then reduces to $\mu(B) \leq \frac{E'}{E} - \mu$ and so
$\int_0^t \mu(B)\, dt < (\log E(t) - \mu t)$

Hence, we finally obtain $\overline{q_2} = \frac{\int_0^t \mu(B)\, dt}{t} < \frac{\log E(t)}{t} - \mu < \frac{\mu}{2} - \mu < 0$ for all
(S(0), E(0), I(0), Q(0)) in the absor bing set, where the bound on the sizes of the
classes are implied by the uniform persistence of the system. The criterion $\overline{q_2} < 0$ is
thus satisfied and so the endemic equilibrium is globally stable. The condition itself
proves the local stability of the endemic equilibrium, as has been shown by Li and
Muldowney [13].

The qualitative aspects discussed in this section highlight the asymptotic behavior
of the system, which provide useful insight on the possibility of developing control
measures to check the spread of malicious attacks. A proper physical fabrication of
the abstractions represented in the modeling process can go a long way in possibly
eradicating or at least in minimizing the impact of most attacks in technological
networks. In the next section, the obtained results will be verified using numerical
simulations.

4 Numerical Methods and Simulation

The qualitative aspects discussed in the previous section will be verified in this section
to establish the basic conditions that decide the nature of the epidemic event caused
by a malicious attack in a technological network. In Fig. 11a, the local stability of
the infection free equilibrium is shown in the susceptible—infectious plane, for three
different values of R_0, each of which satisfies the condition $R_0 \leq 1$, necessary for the
stability of the equilibrium. In each of the three cases, it is observed that the infection

(a)   **(b)**

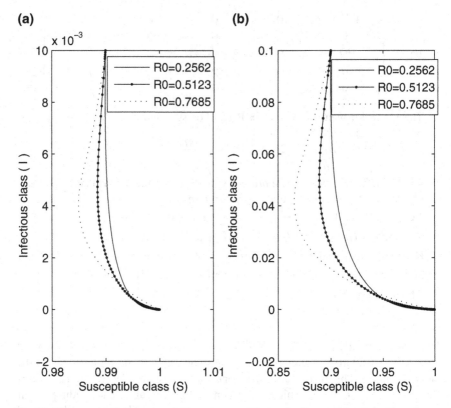

Fig. 11 The asymptotic behavior of the system considering the Susceptible—Infectious plane **a** Local stability **b** Global stability of infection free equilibrium point

free equilibrium is the asymptotically reached point of stability. In Fig. 11b, the global stability has been depicted, for the same set of values of R_0. The observed nature is similar, and asymptotically the state of infection free equilibrium is reached in all of the cases considered. The parameter values are depicted in Table 3. A physical interpretation of the graphs show that the number of nodes in the infectious class ultimately vanishes when $R_0 \leq 1$, whether the infectious fraction is initially itself small (Fig. 11a) or has a much larger value (Fig. 11b). This suggests that if a strong defence mechanism is maintained then the initial level of infection does not have considerable impact on the asymptotic state of the system.

In Fig. 12, the condition for the global stability of the endemic equilibrium is verified, with a value of $R_0 = 1.5370$, which being greater than unity guarantees the global stability of the endemic equilibrium. The behavior is depicted using both the susceptible—infectious plane and the susceptible—exposed plane. In Fig. 12a, the global dynamics of the susceptible—infectious plane is shown, where the system is seen to stabilize at a positive endemic equilibrium, which shows that the infection persists in the population asymptotically. In Fig. 12b, the same phenomenon is

(a)   **(b)**

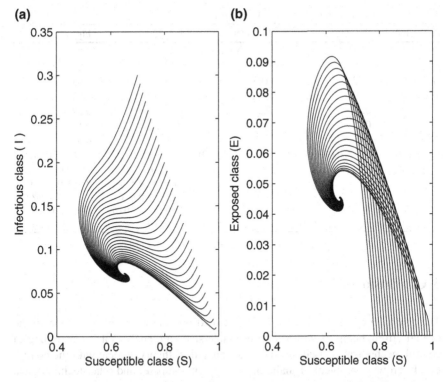

Fig. 12 Global stability of endemic equilibrium shown in **a** S–I phase plane **b** S–E phase plane

observed in the susceptible—exposed plane, where once again the system stabilizes
to the endemic equilibrium point. Physically this means that the number of infected
nodes in the network retains a positive value (infectious fraction of 0.07 and exposed
fraction of 0.045 as shown in Fig. 12a and b respectively).

Numerical simulations can also be used to explore a number of other quantitative
features of the model, including impact of quarantining, the effect of the presence
of latent class of infection and a number of other such properties. For instance, in
the three curves of Fig. 11, the impact of increasing the infectivity contact rate β
(the values considered are 0.05, 0.10 and 0.15) can be easily observed. For a smaller
value of β, the infection is seen to rapidly disappear while for a larger value, the
fall is more gradual. In Fig. 12, the value of β is further increased to 0.30, keeping
all other parameter values constant and it is observed that now the infection persists
in the network. The rate β physically shows the strength of the attacking malicious
object with respect to both its infecting power and also its scanning power to find
new vulnerable hosts in the network.

Table 3 Parameter values for the Figs. 11a, b and 12a, b

Parameter	Fig. 11a	Fig. 11b	Fig. 12a	Fig. 12b
β	(0.05, 0.1, 0.15)	(0.05, 0.1, 0.15)	0.3	0.3
μ	0.04	0.04	0.04	0.04
δ_1	0.03	0.03	0.03	0.03
δ_2	0.02	0.02	0.02	0.02
ε	0.27	0.27	0.27	0.27
γ	0.1	0.1	0.1	0.1
ν	0.06	0.06	0.06	0.06
$(S_0, E_0, I_0, Q_0, R_0)$	(0.99, 0.01, 0, 0, 0)	(0.9, 0.1, 0, 0, 0)	(0.99, 0, 0.01, 0, 0) to (0.70, 0, 0.3, 0, 0) step size (−0.01, 0, +0.01, 0, 0)	(0.99, 0, 0.01, 0, 0) to (0.80, 0, 0.2, 0, 0) step size (−0.01, 0, +0.01, 0, 0)

5 Conclusion

The role of e-epidemic models is expected to provide a newer dimension to the overall efforts being made for a proper understanding, analysis and developing of control mechanisms, based on a concrete analytical theory, to deal with the continuously growing problem of malicious attacks in computer and other technological networks. In this chapter, several of the challenges in applying epidemic modeling to the domain of cyber attacks have been highlighted, which are expected to provide a basic framework that can be utilized while developing further models. The domain is expected to remain challenging in the near future because of the enormous complexities involved, but a proper design and analysis of the domain can lead to comprehensive models, that can possibly help in developing intelligent defense mechanisms that use the analytical framework provided by the models.

References

1. Anderson, R.M., May R M.: Infectious diseases of humans: dynamics and control. Oxford Science Publications, Oxford (2008)
2. Kermack, W.O., McKendrick, A.G.: Contributions to the mathematical theory of epidemics. Proc. R. Soc. A **115**, 700–721 (1927)
3. Kephart, J.O., White, S.R.: Directed-graph epidemiological models of computer viruses. In: Proceedings IEEE ComputerSociety Symposium on Research in Security and Privacy, pp. 343–359, (1991)
4. Mishra, B.K., Saini, D.K.: Seirs epidemic model with delay for transmission of malicious objects in computer network. Appl. Math. Comput. **188**, 1476–1482 (2007)
5. Mishra, B.K., Jha, N.: Seiqrs model for the transmission of malicious objects in computer network. Appl. Math. Modell. **34**, 710–715 (2010)

6. Hyman, J.H., Li, J.: The reproductive number for an hiv model with differential infectivity and staged progression. Linear Algebra Appl. **398**, 101–116 (2005)
7. Hyman, J.H., Li, J.: Differential susceptibility and infectivity epidemic models. Math Biosci. Eng. **3**(1), 89–100 (2006)
8. Mishra, B. K., Ansari, G. M.:Differential epidemic model of virus and worms in compuer network. Int. J. Network Security. **14**(3), 149–155 (2012)
9. Wahl, L.M., Heffernan, J.M., Smith, R.J.: Perspectives on the basic reproductive ratio. J. R. Soc. Interface **2**, 281–293 (2005)
10. Cownden, D., Hurford, A., Day, T.: Next-generation tools for evolutionary invasion analyses. J. R. Soc. Interface **7**, 561–571 (2009)
11. van den Driessche, P., Watmough, J.: Reproduction numbers and sub-threshold endemic equilibria for compartmental models of disease transmission. Math. Biosci. **180**, 29–48 (2002)
12. Wang, L., Li, M.Y., Smith, H.L.: Global dynamics of an seir epidemic model with vertical transmission. SIAM J. Appl. Math. **62**, 58–69 (2001)
13. Li, M.Y., Muldowney, J.S.: A geometric approach to global stability problems. SIAM J. Math. Anal. **27**, 1070–1083 (1996)
14. Wang, L., Li, M.Y.: A criterion for stability of matrices. J. Math. Anal. Appl. **225**, 249–264 (1998)

Chapter 10
Modelling the Joint Effect of Social Determinants and Peers on Obesity Among Canadian Adults

Philippe J. Giabbanelli, Piper J. Jackson and Diane T. Finegood

Abstract A novel framework for modelling trends in obesity is presented. The framework, integrating both Fuzzy Cognitive Maps (FCMs) and social networks, is applied to the problem of obesity prevention using knowledge shared through social connections. The capability of FCMs to handle a large number of relevant factors is used here to preserve domain expertise in the model. Model details and design decisions are presented along with results that suggest that the type of social network structure impacts the effectiveness of knowledge transfer.

1 Introduction

In Canada, one-third of adults are overweight and one-fourth are obese [1], which is similar to the situation in the United States [2]. This high prevalence together with its consequences on population health and the associated treatment costs has

Research funded by the Canadian Institutes of Health Research (MT-10574). We thank the MoCSSy Program for providing facilities..

P. J. Giabbanelli (✉) · P. J. Jackson
MoCSSy Program, Interdisciplinary Research in the Mathematical and Computational Sciences (IRMACS) Centre, Simon Fraser University, Burnaby, Canada
e-mail: giabba@sfu.ca

P. J. Giabbanelli · D. T. Finegood
Department of Kinesiology and Biomedical Physiology, Simon Fraser University, Burnaby, Canada

D. T. Finegood
e-mail: finegood@sfu.ca

P. J. Jackson
School of Computing Science, Simon Fraser University, Burnaby, Canada
e-mail: pjj@sfu.ca

V. Dabbaghian and V. K. Mago (eds.), *Theories and Simulations of Complex Social Systems*, Intelligent Systems Reference Library 52,
DOI: 10.1007/978-3-642-39149-1_10, © Springer-Verlag Berlin Heidelberg 2014

made obesity one of today's most pressing health issues. Far from improving, "the current economic crisis is likely to intensify the obesity epidemic [as we] have fewer options for care available and as healthier, high-cost foods become increasing unaffordable" [3]. We are thus at a time where new approaches should be sought. An approach that is increasingly being advocated is to recognize that obesity is a complex problem [4, 5], impacted by factors such as the psycho-social context (e.g., gender, depression) and the behaviour of peers. These two aspects have been treated extensively but independently. The aim of the model presented in this paper is to foster our understanding of the interplay between these key contributors to obesity.

The importance of peers in obesity was popularized [6, 7] due to an article from Christakis and Fowler [8]. While the methodology was later criticized, the key idea that one's weight is affected by the activities done with peers is well validated [9] and seen as a significant factor [10]. Several models have been designed to investigate this phenomenon. These include an early cellular automaton model [11] later improved to a social network model that accounted for the wide distribution in the number of friends [12]. Both used simple rules stating that weight "spreads" from person to person. We later proposed the first model to study the interplay between behaviour and contextual factors by accounting for the fact that individuals influence activities and not weight directly [13, 14]; this approach has since been advocated by others [15, 16]. However, the context was still crudely defined, and simulation results have pointed to the need for modelling the array of factors that compose it in order to better understand how they interact with the influences conveyed by peers. Studies on the psycho-social factors composing one's psycho-social context abound [17] and have been used to create expert systems [18], but a model has not yet been designed that combines such systems with peer influences. The present paper aims at achieving this combination, based on a new mathematical framework that we developed to accurately represent an individual's context and depict how parts of it change during interactions with others [19, 20].

1.1 Contribution of the Paper

The principal contributions of the present work can be summarized as follows:

- We demonstrate the potential of a new modelling framework for complex social problems by applying it to obesity.
- While previous computational studies of obesity and social networks often focused on harmful effects (e.g., the increased probability to be obese if one has obese peers), we investigate whether leveraging the power of social networks can contribute to mitigating the obesity epidemic.
- We find that the ways in which people are normally connected can limit the efficiency of an intervention aimed at improving knowledge about food and exercise.

1.2 Organization of the Paper

The mathematical framework used to develop our model is presented in Sect. 2, first intuitively and then mathematically; its full mathematical specification can be found in [19] and an illustration is provided in [20]. Section 3 introduces the model built on this framework in a step-by-step approach. In Sect. 4, we use this model to conduct simulations aiming at ameliorating obesity in the population. Results are discussed in Sect. 5 together with possible improvements to the model.

2 Modelling Framework

2.1 Informal Overview

One's obesity is shaped by an environment made of many interacting psycho-social factors. The complexity of obesity is not a mere consequence of the number of these factors or the amount of their interactions. A key challenge is to be found in the *nature* of these interactions: they can be vague, difficult to interpret, or uncertain. Fuzzy Cognitive Maps (FCMs) have been used to address the complex social situations in which such types of interactions are encountered, ranging from diabetes among Canadian aboriginals [21] to homelessness [22] and international politics [23, 24]. FCMs have a network structure: they are composed of nodes representing the factors (e.g., obesity, discrimination) connected by directed edges standing for causal relationships (e.g., impact of obesity on discrimination). An in-depth technical description can be found in [25]. In an FCM, the strength of an edge is assigned by collecting experts' opinions about that edge and combining them using Fuzzy Logic Theory, which is a technique of choice since it "resembles human reasoning under approximate information and inaccurate data to generate decisions under uncertain environments. It is designed to mathematically represent uncertainty and vagueness, and to provide formalized tools for dealing with imprecision in real-world problems [26]". Once all edges have been assigned a strength based on that process, a "case" (e.g., the description of a patient) can be built by assigning a value to each factor of the FCM. Then, a simulation consists of repeatedly updating the values of the factors until a subset of them (e.g., the "obesity" factor) stabilizes.

Fuzzy cognitive maps allow to represent the context in a way that is meaningful both mathematically and in terms of domain expert understanding, which makes it valuable for interdisciplinary endeavours. Furthermore, FCMs are able to include many relevant aspects, that is, aspects considered to be important and capable of being modelled under current conditions. This allows us to avoid the premature elimination of factors that may later turn out to be important. This modelling point is discussed in [27] and the underlying logic against simplicity as a primary modelling heuristic can be found in [28].

Typically, one FCM is used and presented with different cases for which it makes predictions. This framework differs by using one FCM to represent the environment *of each* individual. The *structure* of the FCM is the same for all individuals since that represents causalities, but the *values* of the factors are specific to the individual. For example, Fig. 1 shows three individuals, each with an FCM. We know that obesity generates a weight stigma, and this link is found in all individuals. However, it will only have an effect for Lilian and Garnet, who are the ones experiencing obesity.

The key idea of this framework is that individuals influence and are influenced by others on *some* factors of their FCMs, which are then updated to account for how the environment mediates these social influences. For example, in Fig. 1, Martha might influence her friend Lilian by promoting exercise, thus the value of Lilian's exercise might initially go up. Then, Lilian's FCM is updated, and her socio-economic status will start to mediate Martha's influence, since not being able to afford exercise might not allow Lilian to do the changes advocated by her friend. Overall, this framework can model both a realistic population structure, generated using scale-free or small-world networks, and an accurate environment thanks to a finely-tuned expert system.

2.2 Core Formal Specification

The population is formalized as a set $\mathbb{V}$ of peers (blue circles in Fig. 1) linked by social ties $\mathbb{E}$. Each individual $v \in \mathbb{V}$ has a Fuzzy Cognitive Map, denoted by v_{FCM}. A Fuzzy Cognitive Map is formally defined as follows. Consider a set F of n factors, the matrix $M_{i,j}, i = 1 \cdots n, j = 1 \cdots n$ that encodes the weight of the causal relationship from factor i to j, and the vector of factors' values $V_i(t), i = 1 \cdots n$ at time t. Then, the new values of the factors at time $t + 1$ for a given FCM are obtained by the standard equation

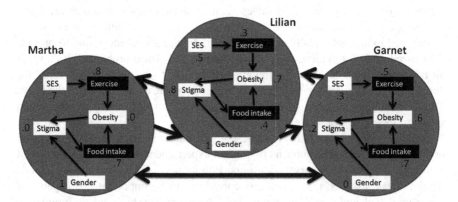

Fig. 1 The initialization provides each individual with an instance of the FCM. Some concepts of the FCM can be influenced by peers (in *black*) while others (in *white*) cannot. This example shows that one's exercise and food intake can be influenced by peers, whereas the socio-economic status (SES) is not perceived as being influenced by peers. The simulation will evolve the influenced factors based on peers' FCMs, and then apply the inference engine to evolve each FCM

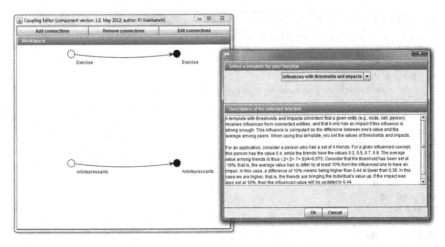

Fig. 2 Using thresholds and impacts, only significantly different peer behaviours promote have an impact, and this impact changes one's current behaviour by a given fraction. This promotes smooth changes, in opposition to adopting extreme behaviour, and was used previously for obesity in [14]

$$V_i(t+1) = f\left(V_i(t) + \sum_{j=1,j\neq i} V_j(t) \times M_{ji}\right) \quad (1)$$

where f is a function that bounds the output in the range $[-1, 1]$, such as *tanh*. The *structure* of the FCM is the same for all individuals, that is, the factors F and their connections $M_{i,j}$ are identical across the population. However, the *value* of these factors at a given time $V_i(t)$ are specific to an individual. Some of these factors are being influenc*ed* by peers and are denoted by $\alpha(F) \subseteq F$. Similarly, the factors that are influenc*ing* peers are denoted by $\beta(F) \subseteq F$. Note that this formalism makes two key assumptions. First, whether a factor is influencing or influenced does not depend on time, as this role is assumed to be constant. Second, influencing and influenced factors are assumed to be the same for all peers, that is, $\forall a, b \in \mathbb{V}, \alpha(a_F) = \alpha(b_F), \beta(a_F) = \beta(b_F)$. These assumptions are the most straight-forward and fit the domain knowledge at a level appropriate for the current modelling goals of our project. The function $S : (a \in \alpha(F)) \mapsto \{b \in \beta(F)\}$ determines, for each influenced concept, the set of concepts that influence it. The key function that determines how the value of a concept changes based on peers' influence is denoted by $\gamma : (\mathbb{R}, \mathbb{R}, \mathbb{R}) \mapsto \mathbb{R}$ and has access to the values of both the influenced and influencing concept, as well as the weight of the relationship. The implementation of this function is scenario specific, and our implementation of this framework currently offers several options. These include adopting the most extreme behaviour, and adopting part of a peer's behaviour if the difference is stronger than a given threshold (Fig. 2). Finally, the evolution of the population for one time step is summarized byAlgorithm 1, which has been further discusse in [19].

Algorithm 1 Evolves the population

Require: The population has been initialized, the FCMs have starting values
1: //*Applies the social influences in parallel*
2: **for** $i \in \mathbb{V}$ **do**
3:   **for** $j \in \mathbb{V}|(j, i) \in \mathbb{E}$ **do**
4:     //*for each neighbor j influencing a person i*
5:     **for** $a \in \alpha(i_F)$ **do**
6:       //*for each concept influenced by peers*
7:       **for** $b \in S(a)$ **do**
8:         //*for each influencing concept*
9:         $V_a(t + 1) \leftarrow V_a(t) + \gamma(V_a(t), V_b(t), W((a, b)))$ //*updates the value of concept a*
10:       **end for**
11:     **end for**
12:   **end for**
13: **end for**
14: //*Evolves each individual's FCM until it stabilizes*
15: **for** $i \in \mathbb{V}$ **do**
16:   **while** i_{FCM} does not stabilize **do**
17:     **for** $g \in i_{FCM}$ **do**
18:       $M_g(t + 1) = f(M_g(t + 1) + \sum_{h \in g_{FCM}, h \neq g} M_h(t) \times W_{h,g})$ //*updates each concept*
19:     **end for**
20:   **end while**
21: **end for**

3 Model of Obesity Among Canadian Adults

Our model is introduced using the same bottom-up approch as applied to the framework in Sect. 2. We start by defining how selected social factors interact to shape obesity in the population (Sect. 3.1). We then detail the probability distributions associated with each factor, which allow us to generate virtual individuals (Sect. 3.2). Finally, we address how individuals influence each other (Sect 3.3).

3.1 Structure of the Fuzzy Cognitive Map

Obesity is a complex problem shaped by the interaction of myriad factors. The Foresight Obesity Map arose from efforts aimed at articulating many of these psychological and sociological factors [29], but it was solely a conceptual model. This prompted us to go further in designing a mathematical model, first using system dynamics [30] and later as a Fuzzy Cognitive Map [18]. The structure of that map was based on a literature review. In it, each edge represents a causation, whose strength was independently assessed by seven international experts on obesity, and those assessments were combined using various fuzzy logic techniques. Validation of this map produced robust predictions consistent with expectations for all generated test cases, regardless of the fuzzy logic technique chosen. However, in much the same way the Foresight Obesity Map had to be reduced for the sake of analysis [31], data is not available to satisfy all of the concepts contained in the map. Therefore, we use a reduced version of the FCM, which only includes concepts that can be operationalized given current data. The concept *psychosocial barriers* was importance in connecting demographic variables to exercise, but no standard index exists to measure it in North America. Therefore, the concept was deleted but its edges were rewired by directly connecting the demographic variables to exercise.

The new hypothesis explored in this paper is the addition of *knowledge*, which stands for the elements that affect an individual's choice regarding the two most promixal factors of obesity: physical exercise and food intake. Improving knowledge regarding food (i.e., nutrition knowledge) could be a valuable component of programs aiming at managing weight, since nutrition knowledge acts through pathways that include "adoption of more healthful cooking methods, improving skill at label reading and meal planning, and reduced consumption of high-energy and high-fat foods" ([32] and references within). The association between knowledge and food intake differs widely between studies [33], thus we take a neutral stand by assigning it an intermediate "medium" value in the FCM. Similarly, knowledge can impact exercise through several means: individuals might not exercise because they do not know about the opportunities available in their neighborhood, which is studied in urban health as "perceptions of the built environment" [34], or they might not know how different types of exercise impact their health, which has been particularly studied among the elderly [35]. Again, the broad definition of knowledge makes it difficult to precisely estimate its impact on exercise, and as with nutrition knowledge we assign it a medium value.

The FCM resulting from the simplification process and the addition of knowledge is represented in Fig. 3 as viewed in the simulation software that implements the framework summarized in Sect. 2.

3.2 Initial Values of the Fuzzy Cognitive Map

Age uses the age pyramid from Statistics Canada, based on the Canadian population as of July 1st 2010 [36]. The pyramid provides data for all Canadians and since we are focusing on adults, its content was scaled to express the fractions of individuals aged 18 to 100+. To assign an age to a virtual individual, fractions are viewed as probabilities, e.g., the fact that 1.66 % of Canadians are aged 18 translates to 1.66 % chance for a virtual individual to be aged 18 (encoded as age = 0). Similarly, **income** uses the 2009 data from Statistics Canada. The consensus on **fatness perceived as negative** and **belief in personal responsibility** is that both are highly prevalent attitudes in North America, as reviewed in [18]. Thus, both are assigned the high value of 0.8. The values for **stress, depression, antidepressants, health,** and **obesity** are drawn from large Canadian datasets based on the (virtual) individual's age (Table 1). The value of **weight discrimination** is set to be that of obesity; in other words, it is assumed that obese individuals experience a prejudice that linearly depends on their obesity. An alternative hypothesis would be to consider that individuals

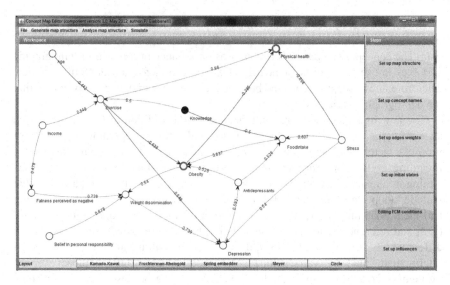

Fig. 3 Fuzzy Cognitive Map of obesity. *Red* edges have negative weights whereas *green* edges have positive weights. Concepts circumferenced in red must stabilize for the FCM to stop evolving, and concepts shaded *black* participate in social influence.

are discriminated against significantly when they are very obese, and negligibly otherwise. However, simulations conducted in [18] for both hypotheses showed that they generally yield almost indistinguishable results.

Regarding **exercise**, we assume as in the United States that most individuals are sedentary [37]. Based on the Food and Agriculture Organization [38], being sedentary translates to a value of 1.53 on the scale known as *level of physical activity* that (in its broader sense) goes from 1.4 to 4.7. Thus, on our scale from 0 to 1, most individuals should have a value of $\frac{1.53-1.4}{4.7-1.4} \approx 0.04$. In order to approximate this heavy-tailed distribution, we use the Inverse Gaussian distribution $f(x; \mu = 1, \lambda = 0.01)$ scaled to the range [0,1], which has a mean of 0.04. Setting **food intake** should not be done independently of the value of exercise. Doing so would result in creating virtual individuals with highly unsustainable and uncommon situations, such as spending significantly more energy than they get. In physiology, food and exercise are connected through the *energy balance*, that is, the difference between intake and expenditure. Precisely estimating this balance in the adult population has been a subject of debates [39, 40], but the recent study by KD Hall and colleagues for the adult American population found that intake was higher than expenditure by 30 kJ on a scale of 11,000. Therefore, on our scale from 0 to 1, we set one's food intake to be higher than exercise by $\frac{30}{11,000} \approx 0.0027$. Finally, as we hypothesized that **knowledge** impacts exercise and food intake, which both have a heavy-tailed distribution, we also assign to knowledge the Inverse Gaussian distribution $f(x; \mu = 1, \lambda = 0.01)$ scaled to the range [0,1].

Table 1 Prevalence of concepts per age bracket from the National Population Health Survey (NPHS) [41], the Canadian Community Health Survey (CCHS) [42, 43], and the Prince Edward Island Nutrition Survey (PEINS) [44]

Measured concept	Source	Sample size	Date	Age cat.	Prevalence (treated here as probability)%
Depressed	NPHS	14,500	1994–1995	8–29	7
				30–39	6
				40–49	6
				50–59	5
				60–69	2
				70+	3
Stressed	NPHS	14,500	1994–1995	18-29	17
				30–39	15
				40–49	16
				50–59	14
				60–69	15
				70+	17
Antidepressants	CCHS	36,984	2002	18–25	Taking antidepressants in the past year: 3.4
				26–45	Taking antidepressants in the past year: 6.3
				46–64	Taking antidepressants in the past year: 7.7
				65+	Taking antidepressants in the past year: 4.1
Obesity	PEINS	1,995	2000	18–34	None:41.7, Overweight:33.8, Obese:24.5
				35–49	None:30.9, Overweight:36.8, Obese:32.3
				50–64	None:22.3, Overweight:41.5, Obese:36.2
				65–74	None:26.8, Overweight:36.3, Obese:36.9
Functional health	CCHS	131,486	2009–2010	18–19	Good to full: 84.7
				20–34	Good to full: 88.1
				35–44	Good to full: 85.5
				45–64	Good to full: 79.7
				65+	Good to full: 67

3.3 Connections Among Fuzzy Cognitive Maps

As highlighted in Fig. 3, we focus on how the concept of *knowledge* influences and is influenced by peers. The function that defines these influences relies on two ideas. First, an individual does not often speak to all of his friends simultaneously. As in [45], for each time step one peer is selected at random for interaction. Second, friends can convey good as well as bad advice, so the direction of influences should not be limited to only improving one's knowledge. Examples of advice that would be detrimental to one's knowledge include recommending inappropriate types of exercise or promoting unhealthy cooking styles. Taking advice from peers is seen as a probabilistic event, whereby one has a probability p of accepting "good" advice (i.e., advice that increases one's knowledge) and $1 - p$ of accepting "bad" advice (i.e., advice that decreases one's knowledge). Formally, denote by $i_{know}(t)$ and $j_{know}(t)$ the value of the concept *knowledge* in individuals i and j at time t. Assuming that j influences i, then $i_{know}(t + 1)$ is given by

$$f(i_{know}(t), j_{know}(t)) = j_{know}(t) \text{ with probability } p \text{ if } j_{know}(t) > i_{know}(t)$$
$$j_{know}(t) \text{ with probability } 1 - p \text{ if } j_{know}(t) < i_{know}(t)$$
$$i_{know}(t) \text{ otherwise}$$

4 Simulation and Results

4.1 Simulation Environment

The model presented in this chapter was developed and tested using CoCo (*C*onnecting *C*oncepts), our software suite designed for building FCM simulations. CoCo is built upon four pillars: a concept map editor (Fig. 3), a coupling editor for designing how nodes can affect each other (Fig. 2), and testing environments for both social and geographic simulations. The concept map editor can be used for experimenting on an individual FCM, which is useful for providing insight or identifying problems before stepping up to a network of FCMs. This model was run using the social simulator, which provides a wide variety of social network models having properties such as small-world or scale-free [46]. The software provides analysis of the graph structure in addition to plots of the system variables over the execution of a simulation run.

4.2 Results

The goal of this simulation is to provide information on how the ways in which individuals are connected (i.e., the structure of the social network) affect *trends* in obesity. Indeed, we are not interested in the *specific* values of obesity, since the effect of knowledge on behaviour change goes through a number of additional steps that have been simplified in the model considered here.

Researchers have found that disparate social networks share numerous properties. Since the first properties were found in the late 1990s, a considerable body of work has been devoted to designing models of networks expressing given properties that are tunable by the user [46]. Two properties have been found to be of particular interest: *small-world* and *scale-free*. Their formal definitions and consequences for various

Table 2 References to network models and parameters used to generate instances

Definition	Parameters	Number of individuals	Small-world property	Scale-free property	Other properties
[49]	$n = 2400$ nodes, edge probability $p = 0.01$	2400	No	No	
[50, 51]	$t = 3$ iterations, $d = 5$ parallel paths	2412	No	Yes	Planar, unclustered
[14, 52]	$n = 185$ base nodes, $\delta = 11$ expected degree	2405	Yes	No	Degree regular
[53]	$t = 4$ iterations, $n = 7$ base nodes	2401	Yes	Yes	

processes can be found in [47, 48]. Informally, the small-world property requires (1) that individuals often belong to communities, and (2) that going from one community to the other requires a small set of intermediate individuals. The scale-free property states that many individuals are linked to a few and a few are linked to many (i.e., the distribution of the percentage of individuals having a given number of friends follows a power-law). In this work, we use several network models in order to obtain different combinations of these properties and study how they impact the outcome of the simulation. Table 2 summarizes the parameters used to generate instances, together with references to the models' definitions. The parameters were chosen to calibrate the networks in size, which allows comparison of simulation results across networks.

Simulations were performed for each of the four networks summarized in Table 2. For each network and different value of the probability p to adopt information (see Sect. 3.3), the value of the obesity concept was computed after 20 steps. Figure 4 reports on the average across 10 runs. This figure demonstrates that the way in which individuals are connected has a significant impact on the potential of knowledge sharing. Explanations for the differences in impact are suggested in the next Section.

It should be noted that the modelling framework used here leads to specific types of simulations. The structure of a model such as designed in Sect. 3.1 synthesizes expert knowledge, and its ties to domain research are strengthened by drawing all values from empirical evidence. Overall, the process used to generate the model produces its central validity, and the goal of a simulation is not to assess this validity once more but instead to provide insight into hidden interactions, for example. The importance of working closely on an empirical level to improve understanding of specific phenomena is emphasized in [54].

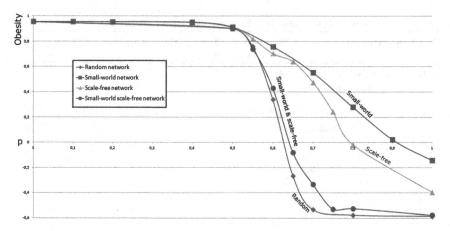

Fig. 4 Average value of the *obesity* concept as a function of the probability to accept a peer's knowledge. $p < .5$ denotes a propensity to accept harmful information whereas $p > .5$ favours beneficial information. *Color online*

5 Discussion

Obesity is one of today's most pressing health issues in countries such as Canada [1]. An individual's obesity is shaped by numerous factors, and this paper focuses on the psycho-social context (e.g., gender, depression) and the effect of peers. While computational models have often treated these two aspects independently, we use a novel framework that is precisely designed to model the interplay between a complex context and peers (Sect. 2). The potential of this framework to foster our understanding of complex social phenomena was demonstrated by building a detailed model of obesity in a step-by-step manner, using intuitive software (Sect. 3). Counter to a number of computational models that have enforced the negative side of peer influence, we use this model to study how individuals can collaboratively lower the prevalence of obesity. This collaborative effort is abstracted as influencing individuals' knowledge on food intake (e.g., learning healthful cooking methods, improving skill at label reading and meal planning) and exercise (e.g., knowing the opportunities available in the neighborhood and the impact of different types of exercise on health). We simulated how the effect of knowledge on obesity was mitigated by several possible ways in which individuals can connect with others (Sect. 4). The key finding of this study is that *any* of the social structures that could take place in the real-world is less efficient at reducing obesity than connecting to random peers. An immediate practical consequence is that the effect of interventions on building knowledge regarding obesity could be limited by the ways in which we naturally befriend others. This effect is particularly salient when observing the relatively low effectiveness of knowledge sharing in the small-world network which, research suggests, is an excellent candidate for the social networks that we naturally create [55].

Both small-world and random networks have a low-average distance, that is, information travels fast. The *one* difference between these two networks lies in communities, which are absent from a random network but essential for a small-world network. Therefore, an explanation is that information travels fast but does not *stick* well, as it is difficult to bring new ideas to a community where individuals reinforce each other in "old habits". Furthermore, our model did not account for the fact that individuals in a community tend to share similar features. This similarity, or *homophily*, is indeed known to "limit people's social worlds in a way that has powerful implications for the information they receive" [56], which here could translate to making the adoption of new knowledge even harder, hence increasing the difference between small-world and random networks.

The model introduced in this chapter can be improved by future work in a number of ways. Firstly, our mathematical framework can bring in homophily by modifying the function representing the influences on knowledge (Sect. 3.3), so that the chance of adopting information is proportional to the similarity between the interacting individuals (e.g., in terms of age or gender), which is akin to the approach undertaken in [45]. Secondly, our simulation allowed us to look at the impact of knowledge on obesity in terms of trends but did not have the accuracy required to predict actual values. This could be improved by extending the model in include stages between knowledge and behaviour change, such as intention. However, the relationships between such stages are often uncertain, as research has shown that intention does not exactly predict changes [57]. An alternative approach could be to focus on a particular population in which the different kinds of information that can be passed on are accurately defined and their impact known. This has been made feasible for Canadian aboriginals through the work of Giles and colleagues who have created a Fuzzy Cognitive Map that can be directly used in our framework and precisely identifies elements such as "disease awareness" or "trust in food" [21].

6 Conclusion

The results presented here which indicate the importance of social network type with regard to knowledge transfer have been generated using a model which is both descriptive and mathematically sound. Findings from further research using our approach could be used to inform public health policymakers; if these results are upheld, then it may be more important that we share good advice about fighting obesity with strangers than with those we know well.

References

1. Public Health Agency of Canada and the Canadian Institute for Health Information: Obesity in Canada (2011)
2. Ogden, C., Carroll, M.: Prevalence of overweight, obesity, and extreme obesity among adults: United-states, trends 1960–1962 through 2007–2008. National Center for Health, Statistics (June 2010)
3. Levi, J., Vinter, S., Richardson, L., Laurent, R.S., Segal, L.: F as in fat: how obesity policies are failing in america. Trust for America's health. Robert Wood Johnson Foundation (July 2008)
4. Finegood, D.: The complex systems science of obesity. In: The oxford handbook of the social science of, obesity, pp. 208–236 (2011)
5. Kumanyika, S., Parker, L., Sim, L., (eds.): Defining the problem: the importance of taking a systems perspective. In: Bridging the evidence gap in obesity prevention: a framework to inform decision making. The national academies press (2010)
6. Kolata, G.: Find yourself packing it on? blame friends. The New York Times (July 2007)
7. Thompson, C.: Are your friends making you fat? The New York Times (September 2009)
8. Christakis, N.A., Fowler, J.H.: The spread of obesity in a large social network over 32 years. N. Engl. J. Med. **357**(4), 370–379 (July 2007)
9. Hammond, R.A.: Social influence and obesity. Curr. Opin. Endocrinol. Diabetes Obes. **17**, 467–471 (2010)
10. Smith, K., Christakis, N.: Social networks and health. Ann. Rev. Soc. **34**, 405–429 (2008)
11. Rush, W., Biltz, G., Pantelidis, L., Kim, L., Ogata, C., Muneichi, S.: An exploration of the influence of health related risk factors using cellular automata. Tech. rep (New England Complex Systems Institute) (2003)
12. Bahr, D., Browning, R., Wyatt, H., Hill, J.: Exploiting social networks to mitigate the obesity epidemic. Obesity **17**(4), 723–728 (2009)
13. Giabbanelli, P.J., Alimadad, A., Dabbaghian, V., Finegood, D.T.: Modeling the influence of social networks and environment on energy balance and obesity. Obes. Rev. **11**(Suppl. 1), 65 (2010)
14. Giabbanelli, P.J., Alimadad, A., Dabbaghian, V., Finegood, D.T.: Modeling the influence of social networks and environment on energy balance and obesity. J. Comput. Sci. **3**, 17–27 (2012)
15. Ruths, D.: Social network modeling of lifestyle behaviors and obesity. In: 2nd National Obesity Summit (2011)
16. Karanfil, O., Moore, T., Finley, P., Brown, T., Zagonel, A., Glass, R.: A multi-scale paradigm to design policy options for obesity prevention: exploring the integration of individual-based modeling and system dynamics. In: Proceedings of the 29th International Conference of the System Dynamics Society (June 2011)
17. Hill, A.: Social and psychological factors in obesity. In: Obesity: Science to Practice, pp. 347–366. Wiley-Blackwell (2009)
18. Giabbanelli, P.J., Tornsney-Weir, T., Mago., V.K.: A fuzzy cognitive map of the psychosocial determinants of obesity. Appl. Soft Syst. 12(12):3711–3724 (2012)
19. Giabbanelli, P.J.: A novel framework for complex networks and chronic diseases. In: Proceedings of the third workshop on complex networks (2012)
20. Pratt, S.F., Giabbanelli, P.J., Jackson, P., Mago, V.K.: Rebel with many causes: a computational model of insurgency. In: Proceedings of the IEEE International Conference on Intelligence and Security Informatics (2012)
21. Giles, B., Findlay, C., Haas, G., LaFrance, B., Laughing, W., Pembleton, S.: Integrating conventional science and aboriginal perspectives on diabetes using fuzzy cognitive maps. Soc Sci Med **64**(3), 562–576 (2007)
22. Mago, V., Morden, H., Fritz, C., Wu, T., Namazi, S., Geranmayeh, P., Chattopadhyay, R., Dabbaghian, V.: Analyzing the impact of social factors on homelessness: A fuzzy cognitive map approach. BMC Med. Inform. Decis. Mak. (2013)

23. Tsadiras, A.K., Kouskouvelis, I., Margaritis, K.G.: Using fuzzy cognitive maps as a decision support system for political decisions. Lecture Notes in Computer Science **2563**, 291–301 (2003)
24. Neocleous, C.C., Schizas, C.N.: Application of fuzzy cognitive maps to the political-economic problem of cyprus. In: Proceedings of the International Conference on Fuzzy Sets and Soft Computing in Economics and, Finance, pp. 340–349 (2004)
25. Glykas, M., (ed.): Fuzzy cognitive maps: advances in theory, methodologies, tools and applications. Studies in fuzziness and soft Computing. Springer (2010)
26. Li, Z.: Introduction. In: Fuzzy chaotic systems, pp. 1–11. Springer (2006)
27. Edmonds, B., Moss, S.: From kiss to kids—an 'anti-simplistic' modelling approach. Lecture Notes in Computer Science **3415**, 130–144 (2005)
28. Edmonds, B.: Simplicity is not truth-indicative, pp. 02–99. CPM, Report (2002)
29. Vandenbroeck, I., Goossens, J., Clemens, M.: Foresight tackling obesities: future choices-building the obesity system map. Government Office for Science, UK Governments Foresight Programme (2007)
30. Giabbanelli, P.J., Torsney-Weir, T., Finegood, D.T.: Building a system dynamics model of individual energy balance related behaviour. Can. J. Diab. **35**(2), 201 (2011)
31. Finegood, D.T., Merth, T.D., Rutter, H.: Implications of the foresight obesity system map for solutions to childhood obesity. Obesity **18**, S13–S16 (2010)
32. Klohe-Lehman, D.M., Freeland-Graves, J., Anderson, E.R., McDowell, T., Clarke, K.K., Hanss-Nuss, H., Cai, G., Puri, D., Milani, T.J.: Nutritionknowledge is associated with greater weight loss in obese and overweight low-income mothers. J. Am. Diet. Assoc. **106**(1), 65–65 (2006)
33. Wardle, J., Parmenter, K., Waller, J.: Nutrition knowledge and food intake. Appetite **34**, 269–275 (2000)
34. McGinn, A.P., Evenson, K.R., Herring, A.H., Huston, S.L., Rodriguez, D.A.: Exploring associations between physical activity and perceived and objective measures of the built environment. J. Pub. Health **84**(2), 162–184 (2007)
35. Schutzer, K.A., Graves, B.S.: Barriers and motivations to exercise in older adults. Preven. Med. **39**(5), 1056–1061 (2004)
36. Statistics Canada: annual demographic estimates: Canada, provinces and territories. Catalogue (91–215-X) (2010)
37. Booth, F., Chakravarthy, M.: Cost and consequences of sedentary living: New battleground for an old enemy. Presidents Council on Physical Fitness and Sports Research Digest **3**(16) (2002).
38. Food and agriculture organization: human energy requirements. Report, joint FAO/WHO/UNU expert consultation (2004)
39. Bouchard, C.: The magnitude of the energy imbalance in obesity is generally underestimated. Int. J. Obes. **32**, 879–880 (2008)
40. Swinburn, B.A., Sacks, G., Lo, S.K., Westerterp, K.R., Rush, E.C., Rosenbaum, M., Luke, A., Schoeller, D.A., DeLany, J.P., Butte, N.F., Ravussin, E.: Estimating the changes in energy flux that characterize the rise in obesity prevalence. Am J Clin. Nutr. **89**, 1723–1728 (2009)
41. Stephens, T., Dulberg, C., Joubert, N.: La sante mentale de la population canadienne: une analyse exhaustive. Maladies chroniques au Canada **20**(3), 131–140 (1999)
42. Beck, C.A., Patten, S.B., Williams, J.V., Wang, J.L., Currie, S.R., Maxwell, C.J., El-Guebaly, N.: Antidepressant utilization in canada. Soc Psychiatry Psychiatr Epidemiol **40**, 799–807 (2005)
43. Statistics Canada: Health indicator profile, two year period estimates, by age group and sex, canada, provinces, territories, health regions, (2011 boundaries) and peer groups. CANSIM Table 105–0502, (2011)
44. MacLellan, D.L., Taylor, J.P., Til, L.V., Sweet, L.: Measured weights in pei adults reveal higher than expected obesity rates. Revue canadienne de sante publique **95**(3), 174–178 (2004)
45. Axelrod, R.: The dissemination of culture: a model with local convergence and global polarization. J. Conflict Resolut. **41**(2), 203–226 (1997)

46. Giabbanelli, P.J., Peters, J.G.: Complex networks and epidemics. Technique et Science Informatiques **30**, 181–212 (2011)
47. Newman, M.E.J.: The structure and function of complex networks. SIAM Rev. **45**(2), 167–256 (2003)
48. Li, L., Alderson, D., Tanaka, R., Doyle, J.C., Willinger, W.: Towards a theory of scale-free graphs: Definition, properties, and implications (extended version) (Oct 2005)
49. Erdos, P., Renyi, A.: On random graphs i. Publicationes Mathematicae **6**, 290–297 (1959)
50. Comellas, F., Miralles, A.: Modeling complex networks with self-similar outerplanar unclustered graphs. Physica A: Stat. Mech. Appl. **388**, 2227–2233 (2009)
51. Giabbanelli, P.: The small-world property in networks growing by active edges. Adv. Complex Syst. **14**(6), 853–869 (2011)
52. Giabbanelli, P.J.: Self-improving immunization policies for complex networks. Masters thesis, Simon Fraser University (2009)
53. Barriere, L., Comellas, F., Dalfo, C.: Deterministic hierarchical networks. Preprint Universitat Politecnica de Catalunya (2006)
54. Boero, R., Squazzoni, F.: Does empirical embeddedness matter? methodological issues on agent-based models for analytical social science. J. Artif. Soc. Soc. Simul. **8**(4) (2005)
55. Schnettler, S.: A structured overview of 50 years of small-world research. Soc. Netw. **21**, 165–178 (2009)
56. McPherson, M., Smith-Lovin, L., Cook, J.M.: Birds of a feather: homophily in social networks. Annu Rev Sociol **27**, 415–444 (2001)
57. Sheeran, P.: Intention-behavior relations: a conceptual and empirical review. Eur Rev. Soc. Psychol. **12**(1), 1–36 (2002)

Chapter 11
Youth Gang Formation: Basic Instinct or Something Else?

Hilary K. Morden, Vijay K. Mago, Ruby Deol, Sara Namazi, Suzanne Wuolle and Vahid Dabbaghian

Abstract As long as people have lived in urban settings, organized criminal gangs have formed. Youth gangs, a special type of organized criminal gang, are made up of predominately male adolescents or young adults who rely on group intimidation and violence. These groups commit criminal acts in order to gain power and recognition, often with the goal of controlling specific types of unlawful activity such as drug distribution. Historically, youth gang formation was attributed to macro-level social characteristics, such as social disorganization and poverty, but recent research has demonstrated a much more complex relationship of interacting factors at the micro-, meso-, and macro-levels. Despite the identification of many of these factors, the journey to gang affiliation is still not well understood. This research, through the application of a fuzzy cognitive map (FCM) model, examines the strength and direction of factors such as early pro-social attitudes, high self-efficacy, religious affiliation, perceptions of poverty (relative deprivation), favorable attitudes towards

We thank the MoCSSy Program for providing financial assistance to the authors and the IRMACS Centre for research facilities.

H. K. Morden (✉) · V. K. Mago · R. Deol · S. Namazi · S. Wuolle · V. Dabbaghian
MoCSSy Program, Interdisciplinary Research in the Mathematical and Computational
Sciences (IRMACS) Centre, Simon Fraser University, Burnaby, Canada
e-mail: hkmorden@gmail.com

V. K. Mago
e-mail: vmago@sfu.ca

R. Deol
e-mail: deol@sfu.ca

S. Namazi
e-mail: sna44@sfu.ca

S. Wuolle
e-mail: srw5@sfu.ca

V. Dabbaghian
e-mail: vdabbagh@sfu.ca

deviance, early onset drug/alcohol use, early onset sexual relations, and their inter-active effects on youth gang formation. FCMs are particularly useful for modeling complex social problems because they are able to demonstrate the interactive and reciprocal factors that affect a given system. Using expert opinion, to determine direction and weight of the influence of the above factors, a FCM was built and validated providing support for the use of FCMs in understanding and analyzing complex social problems such as youth gang formation. This study offers insight into how this type of modeling can be used for policy decision-making.

1 Introduction

Youth gangs and their attendant property crime, drug distribution, and lethal violence cause monetary, social, and personal cost to all who live in North American cities [1–3]. Empirical research has demonstrated that even when controlling for individual level attributes, those who belong to youth gangs commit significantly more crime than those who do not [3–5]. Numerous factors, at the personal, family, school, and community level, have been implicated in the lives of youth who affiliate with gangs [4–6]. Multiple theories have been offered to explain youth gang affiliation including social disorganization (Shaw and McKay 1931, as cited in [7]), low self-control [8], differential association (Sutherland and Cressey 1978, as cited in [7]), subcultural theory (Cohen 1955, as cited in [7]), and strain (Cloward and Ohlin 1960, as cited in [7]). Linked to these factors and theories, thousands of programs in North Amer-ica have been developed and implemented to address this problem; however, today, youth gangs continue to flourish (Gottfredson and Gottfredson 2001, as cited in [5]).

One possible cause for deficiencies in intervention programs may be the data analysis and interpretation used to inform the policies and programs that underlie them. This is because traditional, static, linear, correlational statistical models have been the quantitative tools relied upon to assess data. Youth gang affiliation is a dynamic process and, therefore, if the tools used to measure and assess its impact in a community were also dynamic it is likely that a higher quality of information would be available for policy-making decisions. Models, using FCMs, are ideally suited to dynamic processes, such as youth gang formation, given their ability to capture and represent a large number of interactive and dynamic factors. FCMs are capable of modeling the imprecise nature of human decision-making and allow for multiple, varied experimental conditions in test simulations that result in answers to "what-if" scenarios related to policy-making. Specifically, the answers generated may then be used to help inform decision-making regarding community programs, police strategies, and intervention-based educational programs.

It is generally accepted that a problem is easier and less costly to prevent than it is to solve. This adage can be applied to youth gang affiliation. If youth are prevented from joining or affiliating with gangs and their members then the attendant costs associated with youth gangs may be avoided. However, it must be acknowledged that strategic responses to existing youth gangs must also be attended to and communities must first identify the problem, link the problem to known risk factors, develop strategies that address root causes, and then develop a comprehensive approach for prevention, intervention, as well as the initial suppression of existing gangs [9].

FCMs are a logical and refining step forward from the linear models social scientists have traditionally used to determine correlations between factors related to observed macro-level phenomena such as are seen in youth gangs. FCM models are capable of great complexity; however, for greatest utility and understanding, it is wise to begin with a simple model, using well-studied factors that have been empirically shown to affect the micro-level phenomena. For this reason, we begin with a simple model including factors that both encourage and discourage youth gang affiliation. This type of model provides a basis for hypothesis testing, clarification of theory, and refinement over existing linear statistical models, and helps create a foundation for future, more complex models.

2 Conceptual Model

A simple FCM allows for a baseline model of youth gang affiliation by using well-known factors to test the applicability of this type of model for this type of inquiry. Variables of interest are represented as nodes with links (edges) showing the direction of the interactions of the variables and the resulting effect on the node of interest (gang affiliation). Our model includes eight nodes, empirically shown as representative of some of the most prominent, personal factors related to early onset gang affiliation (gang affiliate and pre-gang affiliate behaviors that occur during late childhood and early adolescence) as well as those identified that act as protective factors. Factors increasing the likelihood of gang affiliation, perception of poverty, favorable attitudes towards deviancy and violence, early onset sexual behavior, and early onset drug/alcohol use, were taken from a larger group of factors commonly implicated by professionals in the movement of youth towards gang affiliation. Factors decreasing the likelihood of gang affiliation, pro-social attitudes in early childhood, high self-efficacy, and religious affiliation, were also drawn from a larger group of factors commonly implicated by professionals in preventing or reducing the likelihood of deviance and subsequent gang affiliation [2, 4, 10]. The central node is level of gang affiliation. The factors used in this FCM were chosen, not because they are an exhaustive list, but because they are a representative list of micro-level factors related to youth gang affiliation. Simple maps, such as this, help establish a foundation from which to build more complex and representative models as well as permit experimentation, the results of which can be used in practical policy applications.

2.1 Factors Implicated in Gang Affiliation

Risk factors are embedded in the five domains of an individuals life: personal characteristics, family conditions, school, peer group, and the community [5]. An accumulation of risk factors greatly increases the likelihood of gang involvement [2]. Some factors are generated as a result of the individual's biology and psychology while others are generated within primary and secondary social contexts. Attitudes and behaviors regarding perceptions of poverty and favorable attitudes towards deviant or early-onset adult behaviors, such as sexual activity and the consumption of drugs and alcohol, have been shown to be commonly present in youth who affiliate with gangs [11].

Perception of Poverty

Poverty or low socio-economic status, in relation to youth who affiliate with gangs, is not an absolute measure of deprivation but a relative measure of deprivation. This is the level of financial deprivation a youth expresses feeling as compared to an absolute measure of their ability to pay for necessities and luxuries [12]. Youth who join gangs were traditionally shown as coming from socially and economically disadvantaged neighbourhoods and homes, but more recent research demonstrates that genuine economic disadvantage has less of an effect than does the perception of economic disadvantage [12].

Favorable Attitudes Towards Deviance and Violence

Youth who demonstrate high levels of dysfunctional and anti-social behavior are at high risk to join gangs [5]. These individual anti-social behaviors often emerge in early childhood [13] and those who show favorable attitudes towards deviance and violence tend to affiliate more often with gangs and commit more violent offenses than youth who do not [2]. In a Denver sample of adolescents, those who demonstrated high tolerance for peer deviance were much more likely to claim gang affiliation than those who showed lower tolerance levels [14]. The statistically significant association between favorable attitudes towards deviance and violence and gang affiliation is considered robust [2]. In Canada, gang members have high self-reported rates of violence, often using violence as the preferred method of solving inter-personal conflict [15].

Early Onset Sexual Behavior

Early dating and precocious sexual activity are strongly related to negative life events and high levels of delinquency [5, 16]. In turn, high levels of delinquency are strongly associated with gang affiliation. Longitudinal studies of childhood risk factors and gang affiliation demonstrate that early sexual activity for adolescents was significantly associated with gang membership [2].

Early Onset Drug/Alcohol Use and Abuse

Youth who affiliate with gangs are often involved in early-onset drug and alcohol use and abuse [16]. In a Canadian cross-country study of adolescent gang members more than 50 % admitted to drug use and approximately 70 % admitted to selling drugs [4].

2.2 Factors Protective Against Gang Affiliation

Protective factors are variables present in childhood and adolescence that diminish the likelihood of negative health and social outcomes [17]. Protective factors help prevent or deter youth from becoming engaged with, and eventually involved in, criminal activity [17]. Protective factors that are the result of biology, such as female gender, good cognitive performance, and lack of learning disabilities are usually present at birth, though may be modified throughout life. Emotional and situational factors such as presence of parental figures at key times during the day (i.e., waking, after school, at dinner, and bedtime), shared activities with these loving, interested adults, including high expectations for behavior and achievement outcomes at school and in extra-curricular activities, are all highly modifiable and have a strong effect on behavior and have been shown as correlated to factors such as pro-social attitudes in early childhood, high self-efficacy, and religious affiliation [17]. These factors were chosen to represent example protective factors that can be modified by contextual inputs and therefore are suitable for testing in a FCM.

Pro-Social Attitudes in Early Childhood

Pro-social attitudes are generally established in early childhood as a result of loving and supportive family environments [8]. Pro-social attitudes are deeply embedded within the family, as an environment and a system, and are represented by the family members' ability to be appropriately responsive to other members' behavior and actions [8, 10, 18]. Strong supportive community and school environments can further strengthen a child's pro-social development [8, 18]. While these attitudes are partially dependent upon the nature of the individual, research has demonstrated that behavior modification is possible and likely in strongly supportive and positive environments [18]. Individuals with strong pro-social beliefs are unlikely to be attracted to deviant behaviors and even more unlikely to become involved in gang activities [8].

High Self-Efficacy

Self-efficacy develops within family efficacy, where the concept of causality and outcome is first learned; a theme found in the developmental theories of Piaget (1954) and Mead (1934) (Gurin and Brim 1984, as cited in [19]). The ability of the adolescent to feel as though their presence, decisions, and behavior may effect change within themselves and their social environment is expressed and experienced from early childhood, through familial interaction, and later through social interactions at school and within their larger community [18]. The ability of the youth to perceive their actions and behavior as consequential, sets the stage for symbolic interactionism during which the developing child and adolescent practice sources of efficacy through vicarious experience, verbal persuasion and emotional arousal [18].

High levels of self-efficacy are important for adolescents given their dynamic physical, cognitive, and emotional selves [20]. Youth who demonstrate high levels of self-efficacy tend to see themselves as both connected to their community and their current selves as well as an increasingly complex potential self [20]. The chaotic nature of gang affiliation is contrary to high levels of efficacy.

Religious Affiliation

Religious affiliation often occurs within the primary environment of the child and helps to form general beliefs about oneself and the world the individual lives within [21]. Religion is often related to levels of social efficacy that are, in turn, directly related to collective efficacy or a general belief that the system of society works as a whole to foster achievement and personal mastery within the sub-domains of the self, the family, schools and any other social sub-groups [1]. Collective efficacy perceptions begin within group members who consider the various sources of information from self, family, school, church, and society and determine the likelihood of success for various pursuits [1]. In a longitudinal study of children, religion was found directly linked to the individuals' pro-social behavior [22].

3 Fuzzy Cognitive Map

This section introduces the background of FCMs and their suitability for complex social problems, followed by the conversion of our conceptual model to a mathematical model using expert opinion. A final, overall model is presented along with experimentation, demonstrating the use of FCMs for practical applications in policy decision-making.

3.1 FCM in Complex Social Problems

The theory of cognitive maps was developed in 1948 [23] and used to demonstrate causal relationships between factors or nodes of complex systems. This was later adapted, through the application of fuzziness, by Kosko in 1986 [24], creating the theory of FCMs. FCMs are advantageous over static descriptive methods due to their ability to dynamically model complex social relationships and systems [25, 26], utilize qualitative opinion through a mathematical conversion process, and provide policy-makers a decision support tool that can answer "what-if" questions regarding the system [27].

In social science, expert opinion often forms the basis of the core data available to describe relationships between sociological or psychological concepts [28]. Uncertainty and vagueness are often found in data related to adolescent gangs due to the vast differences in opinions and theories of researchers as well as the manner in which influential factors are measured [29]. The FCM model is particularly effective for dealing with this kind of uncertainty and variability of factors found in social sciences due to the application of fuzzy logic in the weighting of the edges between nodes [27]. This has prompted researchers to turn to modeling techniques when attempting to more accurately describe and hypothesis-test complex social systems (see, for example fuzzy cognitive maps for decision support in an intelligent intrusion detection system, [30] and using simulation to test criminological theory, [28]).

FCMs have been used to create policy decision-making support systems where experimentation would be unworkable, too expensive, or immoral to conduct. The purpose of the adolescent gang FCM is to guide researchers and policy-makers in finding appropriate preventative and intervention strategies for the reduction of adolescent gangs through a better understanding of the actions and interactions of risk and protective factors which impact adolescent decision-making regarding gang affiliation.

3.2 Construction of the Fuzzy Cognitive Map

For an in-depth description of how fuzzy cognitive maps are constructed please see [31]. In this section we provide an overview of the FCM as it relates to the technical choices and underlying theoretical foundation of adolescent gang affiliation.

FCMs have a network structure that is composed of nodes that represent domain concepts. In this case, these concepts include risk and preventative factors which impact the decision-making of adolescents in regards to gang affiliation. The nodes are connected by directed edges or links that represent the causal relationships. For example, the FCM shows there is a relationship between *religious affiliation* and *early onset sexual behavior*, such that religious affiliation reduces the likelihood of early onset sexual behavior. Edges are weighted to take on a value between negative 1 and positive 1 where the degree of effect of the antecedent node(i) on the consequent node(j) is ascertained through the conversion of qualitative statements to mathematical values. The causal relationship determines the weight of an edge from concept i to j and, in this case, is qualitatively determined through a survey of expert opinion. If the causal relationship is absent, or concept i has no effect on concept j, then the weight of the edge is said to be 0. If concept i increases concept j, then the weight is said to be between 0 and positive 1, depending on the qualitative value assigned by expert opinion. If concept i decreases concept j, then the weight is said to be between 0 and negative 1. Conceptual maps, as shown in Fig. 1, record the direction of the relationship and the effect (positive, null, or negative) that the antecedent node has on the consequent node. These relationships can also be expressed through an adjacency matrix, W.

$$
W = \begin{array}{c} \\ PSA \\ FAD \\ HSE \\ ESB \\ EDA \\ RA \\ PP \\ GA \end{array}
\begin{array}{c} PSA \end{array}
\left(\begin{array}{cccccccc}
& PSA & FAD & HSE & ESB & EDA & RA & PP & GA \\
0 & -0.3716 & 0.6059 & 0 & 0 & 0 & 0 & -0.6664 \\
0 & 0 & 0 & 0.6280 & 0.6664 & -0.2265 & 0 & 0.7727 \\
0 & -0.2724 & 0 & -0.4381 & -0.3926 & 0 & 0 & -0.4984 \\
0 & 0 & 0 & 0 & 0 & 0 & 0 & 0.2943 \\
0 & 0 & 0 & 0 & 0 & 0 & 0 & 0.7727 \\
0 & -0.4381 & 0.3321 & -0.4381 & -0.3331 & 0 & 0 & -0.4984 \\
0 & 0 & 0 & 0 & 0 & 0 & 0 & 0.6059 \\
0 & 0 & 0 & 0 & 0 & 0 & 0 & 0
\end{array} \right)
$$

To collect expert opinion, to weight the edges of the FCM, a survey was designed to assess the strength and direction of the relationships between each pair of concepts of interest (see Appendix A for expert opinion summary). This survey consisted of a list of the relationships with linguistic terms such as "very low", "low", "medium", "high or "very high" and the direction in which the first concept affected the second such as positively (increased), negatively (decreased), or no effect. We presented this survey to a range of experts in youth gangs from Canada and the United States. Perceptions of what constitutes, for example, a "medium" relationship differs amongst individuals. Thus, the conversion of the qualitative statement to mathematical value was allowed to overlap as shown in Fig. 2. Following the fuzzy logic process, developed by Zadeh [32], the imprecise concepts in real-world problems were converted and used in this model.

The combination of expert opinion and overlapping perception can best be understood by considering an illustration of a sample edge and how it is weighted. Consider the impact of *pro-social attitudes in early childhood* on *gang affiliation*. One expert

Fig. 1 Fuzzy cognitive map of gang affiliation

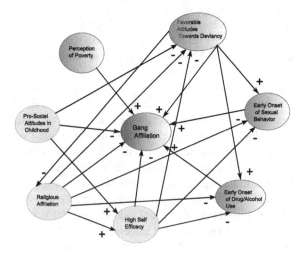

states that this causal relationship is *Very High*, other says *High* and the third expert choose *Medium*. Using these opinions for constructing IF-THEN rules for the fuzzy inference system, we calculate the numeric value. In this example these rules would be listed as follows:

- IF *pro-social attitudes in early childhood* THEN *gang affiliation* is Negatively **Very High** (0.33)
- IF *pro-social attitudes in early childhood* THEN *gang affiliation* is Negatively **High** (0.33)
- IF *pro-social attitudes in early childhood* THEN *gang affiliation* is Negatively **Medium** (0.33)

Summarizing the rules, all experts have different opinion about the strength of association between these two concepts. The set of all edges and all rules forms the knowledge base of this FCM. Values are calculated using the knowledge base and the concepts defined via the linguistic terms. For the above mentioned case see Fig. 3. We used Fuzzy Toolbox from MATLAB R2009b to construct the system.

Figure 3 shows the shaded region of three rules and their output. The shading is based on the number of experts supporting each rule. Outputs, also known as

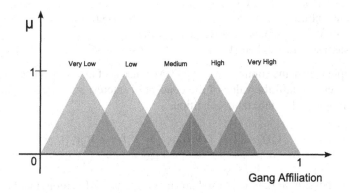

Fig. 2 Fuzzy triangular membership function

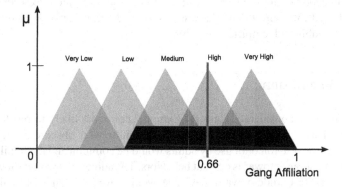

Fig. 3 Deffuzzification of rules using centroid method

"aggregations", can be combined in a variety of ways based on mathematical family functions. Once an aggregation operator has been applied, the final value for that edge is determined through a process known as "centroid defuzzification" using a Mamdani inference mechanism. Using the number of experts supporting each rule is known as the confidence factor and very important in this process as it can significantly affect the edge's weight. This process is used to determine the weight of all edges in the map. (For a further and more thorough description of the process and use for edge weights, please see [33]).

3.3 Fuzzy Cognitive Map Model

Weighted values, shown in matrix W, are placed on the edges and depict the strength of the causal relationship between antecedent and consequent nodes. The node of interest or destination node is gang affiliation.

In order to explain the technical choices, the formalism of FCM is as follows:

- total number of concepts is denoted by n
- the matrix, $W_{ij}, i = 1 \ldots n, j = 1 \ldots n$ denotes the weight of causal relationships from concept i to concept j
- the initial values for each concept are stored in the vector V_i
- the vector V_i is presented iteratively, as per the Eq. 1,
- the system stabilizes when $|V_i(t + 1) - V_i(t)| <= \epsilon$.

In simple words, the simulation of the FCM consists of updating the values stored in V and iterates until the value of the concept of interest stabilizes. In this work, gang affiliation is the stabilizing condition.

$$V_i(t + 1) = f(V_i(t) + \sum_{j=1} V_j(t) \times W_{ji}) \tag{1}$$

where f is a threshold function (also known as a "transfer" function) that bounds the output in the interval $(-1, +1)$. It is common practice to use such a function in order to keep concepts within the specified range [34]. Once weighted, experimentation was conducted to ensure the model was a reasonable representation of reality and agreed with published, empirical literature.

4 Experimentation

Three cases were considered: **Case (I)** Extreme case most likely to result in gang affiliation. This youth is living in a chaotic family with absentee or changing parental figures. Drug/alcohol use by adults would be common and the family would demonstrate tolerance towards deviant behaviors. The adolescent would show exceptionally favorable attitudes towards deviance, would have commenced sexual activity and the consumption of drugs and/or alcohol during early adolescence and would believe that he or she had insufficient money to purchase the luxuries or consumable goods they desired. Stable and low levels of pro-social attitudes would emerge in this environment and such a youth would consider theft as a legitimate way to acquire goods they desired but could not afford. Youth who show early onset of adult behaviors often believe that they do not have control over themselves or their lives and, as a result, have low levels of self-efficacy. The youth and his or her family do not attend church nor hold any spiritual beliefs. During iterations of the FCM pro-social attitudes would remain stable and low. Favorable attitudes towards deviance would initially fall during the affiliation process with the gang as new beliefs, based on the rules and beliefs of the gang emerged to replace any pre-existing pro-social proclivities. Levels of self-efficacy would rapidly fall as the youth came to understand that their decisions had little or no relevance or effect within the gang environment. Any prior religious belief or faith in social conventions would also rapidly fall and remain at very low levels. Gang affiliation, both as a cause and an effect, would rise to a certainty and remain there (Fig. 4).

Case (II) Extreme case least likely to result in gang affiliation. This youth is probably living in a stable, loving family environment where the parental figures set reasonable boundaries on behavior. The family attends church on Sundays and shares many other activities together. The youth would be taught respect for self and others with the parents demonstrating this through moderate consumption of alcohol and a committed, exclusive relationship. This youth would enter elementary school with a strong sense of him or herself, their place in the world and the ability for their decisions to affect their own behavior and others (high self-efficacy). The combination of pro-social attitudes and moderate living would lead to low tolerance for deviance, drug/alcohol use and would also result in an abstention from early onset sexual behavior. During the iterations, pro-social behaviors would remain stable while favorable attitudes towards deviance would rapidly fall and remain low.

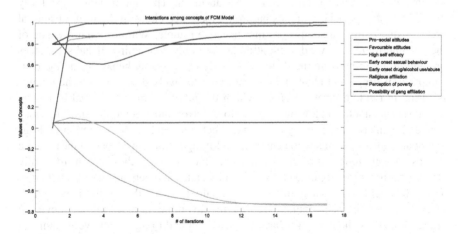

Fig. 4 An experimental case where a person is most likely to affiliate with gangs

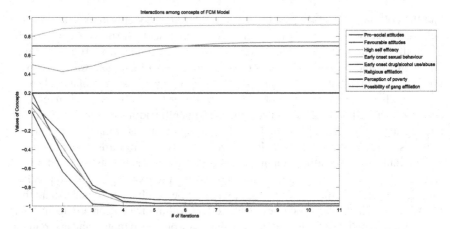

Fig. 5 An experimental case where a person is least likely to affiliate with gangs

Self-efficacy would begin high and rise steadily as the youth experienced the feed-back from their strong pro-social decision-making. The likelihood of early onset sexual behavior or the consumption of drugs/alcohol would begin low and drop to a negligible rate. Religious affiliation would rise as youth who hold strong beliefs in social right and wrong often find this mirrored in the tenets of religious organizations. Perception of poverty would remain stable with little overall influence on behaviors. The likelihood of affiliating with a gang would emerge low and fall to negligible rates during the simulation. In this case the protective properties of pro-social behavior, high self-efficacy and religious affiliation negate any feelings of poverty and remove the likelihood of any forms of deviancy. This combination of factors together and separately would result in no possibility of gang affiliation (Fig. 5).

Case (III) Random case, representative of an average family and real-world adolescent behavior in which gang affiliation is uncertain. This family would have fairly high levels of pro-social attitudes but may experience chaos in the form of marital breakdown, death of a spouse or other negative trauma. Within the family, one of the parental figures shows deviant attitudes, possibly consuming illegal substances or committing minor criminal behaviors. Early onset sexual behavior and the consumption of drugs and alcohol would not be effectively dealt with by the parents resulting in random levels of chaos within the family home. Pro-social behaviors would remain stable and low during iterations. Favorable attitudes towards deviance would remain constant as early onset sexual behavior and the consumption of drugs and/or alcohol are not met with behavior-modifying actions on the part of the parental figures. Despite religious affiliation and self-efficacy rising, these are insufficiently strong to counter the negative effects of family chaos and personal deviant attitudes. While the likelihood of gang affiliation is initially low with the family showing high levels of pro-social behaviours, over time the favorable attitudes towards deviance are reinforced by the deviant behaviours acted out and these would overwhelm any

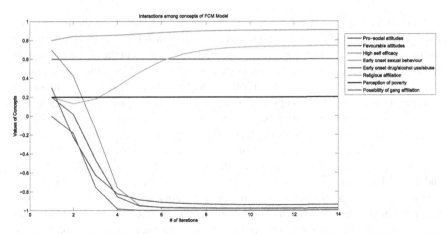

Fig. 6 An experimental case where a person may or may not affiliate with gangs

protective factors the youth may have and gang affiliation becomes more certain (Fig. 6).

5 Discussion and Conclusion

This research helps to confirm the efficacy of using an FCM to model the complex social factors related to gang affiliation by responding in a logical and empirically supported manner. FCMs are particularly suited to this form of experimentation as they provide insight into the interacting effects of negative and positive factors that affect youth today, leading to, or preventing gang affiliation. Micro-level models, such as this, provide a method of looking at the effects of pro- and anti-social behaviors in a way that is difficult in the real world. Models permit the construction of systematic representations of dynamic social interactions by capturing their key components and using them in simulations to determine potential outcomes. Thus, the varying influences of each component and its interaction with other factors, can be examined in a manner that is often impossible, and almost always unethical, in the physical world. This model demonstrates the importance of factors such as self-efficacy and pro-social attitudes acquired within the individual's primary family and could be used to demonstrate to policy-makers that programs which result in strengthening families may be the most cost efficient and effective intervention if the goal is to reduce the societal costs of youth criminal gangs. It is acknowledged that this is a very simple model; however it does demonstrate the efficacy of using FCMs to model this type of social system and future models should certainly be more complex by adding other identified factors implicated in youth gang formation such as those which emerge from school environments (for example negative labeling by teachers and peer group influence) and the larger community environment (for example, community tolerance towards visible gang behavior and towards the availability and use of illegal drugs). FCM models allow for the manipulations of the initial conditions and allow for comparison between models in the pursuit of insight into the likely dynamics of certain combinations of micro-level and meso-level social factors. Models operate bottom-up, allowing an inductive approach to understanding a complex social problem, as opposed to statistical/equation-based explanations that identify correlations between aggregate observations and then attempt to make observations about micro-level mechanisms. FCMs can be used in combination with other forms of models, such as agent-based models, with the goal of allowing researchers to assess the generative sufficiency of a theory by answering the question "Do the individual-level behaviors result in the phenomena described by the theory?". This allows for greater confidence in the validity of a theory to explain youth gang affiliation and help policy-makers to identify key areas most likely to respond favourably to government or agency intervention. FCMs also permit the examination of combinations of factors which result in meeting the threshold necessary to push a youth towards gang affiliation as well as those which do not, providing insight into the combinations of factors most likely to respond to policy changes. The use of fuzzy logic in the FCM extends a greater sense of reality to the modeled behaviors of youth and allows for membership in more than one category at the same time. This avoids the dichotomies that are currently accepted in linear statistical models that limit their applicability to the real world. After all, youth do not suddenly move from the category of non-gang member to gang member in a single, simple step. Research has shown that it is the combinations of factors, in varying strengths, experienced simultaneously or sequentially that ultimately influence the youth towards or away from gang affiliation. FCMs permit the use of a variety of qualitative sources of data, including expert opinion (as used in this case) as well as historical empirical studies, allowing a greater meaning to be given to the vast quantity of literature that has accumulated over the last decades. Fuzzy sets permit the identification of necessary and sufficient conditions for a specific phenomenon to occur [27] representative of the manner in which social science is understood, incorporating ambiguities that are often a part of social science theories [27]. From a study such as this it becomes clear that traditional statistical methods and modeling belong together. The use of empirical theory and prior studies, in which factors have been correlated, informed the choice of factors used in this FCM and allowed for the construction of the surveys provided to experts in the field of youth gang research to rank for relationships and strength of influence. This research offers FCMs as an extension of traditional statistical measurement and inquiry and supports the use of such a model for complex social systems. The basic factors of social science, including ambiguity from the multiple meanings for concepts and the vagueness that occurs from the lack of firm boundaries, are especially suited to the use of FCMs [27]. While FCMs are not a perfect representation of reality they are a refined representation of reality and thus especially useful in policy-making for individuals and governments.

Appendix

Factor A	Factor B	Weight	Weight	Weight	Infl
Pro-social attitudes in early childhood	Gang affiliation	VH	H	M	neg
Pro-social attitudes in early childhood	Youth holding Favorable attitudes towards deviance/violence	VL	M	M	neg
Pro-social attitudes in early childhood	High levels of self efficacy	H	H	M	pos
High self efficacy	Gang affiliation	H	M	L	neg
High self efficacy	Youth holding Favorable attitudes towards deviance/violence	VL	L	L	neg
High self efficacy	Youth engaging in early onset sexual behavior	L	M	M	neg
High self efficacy	Youth engaging in early onset drug/alcohol use/abuse	L	L	M	neg
Religious affiliation	Youth experiencing high self efficacy	L	L	L	pos
Religious affiliation	Gang affiliation	M	H	L	neg
Religious affiliation	Youth engaging in early onset drug/alcohol use/abuse	VL	M	L	neg
Religious affiliation	Youth engaging in early onset sexual behavior	L	M	M	neg
Religious affiliation	Youth holding Favorable attitudes towards deviance/violence	L	M	M	neg
Perception of poverty	Gang affiliation	H	H	M	pos
Favorable attitudes towards deviance/violence	Gang affiliation	VH	VH	H	pos
Favorable attitudes towards deviance/violence	Early onset drug/alcohol use/abuse	VH	H	M	pos
Favorable attitudes towards deviance/violence	Early onset sexual behavior	VH	M	M	pos
Favorable attitudes towards deviance/violence	Religious affiliation	L	VL	VL	neg
Early onset sexual behavior	Gang affiliation	M	VL	VL	pos
Early onset drug/alcohol use/abuse	Gang affiliation	VH	H	VH	pos
#1 Corrections (Parole/Probation) US					
#2 RCMP - undercover					
#3 Gang expert — Toronto, Canada					

References

1. Bania, M.: Gang violence among youth and young adults:(dis) affiliation and the potential for prevention. IPC Rev. **3**, 89–116 (2009)
2. Hill, K., Howell, J., Hawkins, J., Battin-Pearson, S.: Childhood risk factors for adolescent gang membership: results from the seattle social development project. J. Res. Crime Delinquency **36**(3), 300–322 (1999)
3. Tita, G., Ridgeway, G.: The impact of gang formation on local patterns of crime. J. Res. Crime Delinquency **44**(2), 208–237 (2007)
4. Huff, C.: Comparing the criminal behavior of youth gangs and at-risk youths. US Department of Justice, Office of Justice Programs, National Institute of Justice (1998)
5. Wyrick, P., Howell, J.: Strategic risk-based response to youth gangs. Juvenile Justice **9**(1), 20–29 (2004)
6. Hitchcock, H.: Adolescent gang participation: psychological perspectives. J. Police Crim. Psychol. **16**(2), 33–47 (2001)
7. Ball, R., Lilly, J., Cullen, F.: Criminological Theory: Context and Consequences. Sage, Beverly Hills (2010)
8. Hirschi, T.: Causes of delinquency. Transaction Pub (2002)
9. Meacham, M., Stokes, T.: The life development of gang members: interventions at various stages. Forensic Examiner **17**(1), 34–40 (2008)
10. Laub, J., Sampson, R.: Shared Beginnings, Divergent Lives: Delinquent Boys to Age 70. Belknap Press, Cambridge (2003)
11. McBride Murry, V., Berkel, C., Gaylord-Harden, N., Copeland-Linder, N., Nation, M.: Neighborhood poverty and adolescent development. J. Res. Adolesc. **21**(1), 114–128 (2011)
12. Kawachi, I., Kennedy, B., Wilkinson, R.: Crime: social disorganization and relative deprivation. Social Sci. Med. **48**(6), 719–731 (1999)
13. Howell, J.: Youth gangs: prevention and intervention. In: Murray, C., Greenberg, M (eds.) Intervention with Children and Adolescents: An Interdisciplinary, Perspective, pp. 493–514. Allyn and Bacon, Boston (2003)
14. Esbensen, F., Huizinga, D.: Gangs, drugs, and delinquency in a survey of urban youth. Criminology **31**(4), 565–589 (2006)
15. Wortley, S., Tanner, J.: Social groups or criminal organizations? the extent and nature of youth gang activity in toronto. In: From Enforcement and Prevention to Civic Engagement: Research on Community Safety, pp. 59–80. University of Toronto, Toronto (2004)
16. Thornberry, T., Krohn, M., Lizotte, A., Smith, C., Tobin, K.: Gangs and delinquency in developmental perspective. Cambridge University Press, Cambridge (2002)
17. Resnick, M., Bearman, P., Blum, R., Bauman, K., Harris, K., Jones, J., Tabor, J., Beuhring, T., Sieving, R., Shew, M., et al.: Protecting adolescents from harm. JAMA **278**(10), 823–832 (1997)
18. Bandura, A.: Self-efficacy: toward a unifying theory of behavioral change. Psychol. Rev. **84**(2), 191–215 (1977)
19. Gecas, V.: The social psychology of self-efficacy. Annu. Rev. Sociol. **15**, 291–316 (1989)
20. Arnett, J.: Adolescence and Emerging Adulthood. Pearson Prentice Hall, Upper Saddle River (2004)
21. Bandura, A.: On the psychosocial impact and mechanisms of spiritual modeling. Int. J. Psychol. Relig. **13**(3), 167–173 (2003)
22. Bartkowski, J.P., Xu, X., Levin, M.L.: Religion and child development: evidence from the early childhood longitudinal study. Soc. Sci. Res. **37**(1), 18–36 (2008)
23. Tolman, E.: Cognitive maps in rats and men. Psychol. Rev. **55**(4), 189–208 (1948)
24. Kosko, B.: Fuzzy cognitive maps. Int. J. Man-Mach. Stud. **24**(1), 65–75 (1986)
25. Giabbanelli, P.J., Torsney-Weir, T., Mago, V.K.: A fuzzy cognitive map of the psychosocial determinants of obesity. Appl. Soft Comput. **12**, 3711–3724 (2012)

26. Pratt, S., Giabbanelli, P., Jackson, P., Mago, V.: Rebel with many causes: a computational model of insurgency. In: IEEE International Conference on Intelligence and Security Informatics (ISI), IEEE 2012, pp. 90–95 (2012)
27. Ragin, C., Pennings, P.: Fuzzy sets and social research. Sociol. Methods Res. **33**(4), 423–430 (2005)
28. Birks, D., Townsley, M., Stewart, A.: Generative explanations of crime: using simulation to test criminological theory. Criminology **50**(1), 221–254 (2012)
29. Sharkey, J., Shekhtmeyster, Z., Chavez-Lopez, L., Norris, E., Sass, L.: The protective influence of gangs: can schools compensate? Aggression Violent Behav. **16**(1), 45–54 (2011)
30. Siraj, A., Bridges, S., Vaughn, R.: Fuzzy cognitive maps for decision support in an intelligent intrusion detection system. In: IFSA World Congress and 20th NAFIPS International Conference: Joint 9th. vol. 4. IEEE 2001, pp. 2165–2170 (2001)
31. Mago, V.K., Bakker, L., Papageorgiou, E.I., Alimadad, A., Borwein, P., Dabbaghian, V.: Fuzzy cognitive maps and cellular automata: an evolutionary approach for social systems modelling. Appl. Soft Comput. **12** (2012), 3771–3784
32. Zadeh, L.: Fuzzy sets. Inform. Control **8**(3), 338–353 (1965)
33. Mago, V., Mehta, R., Woolrych, R., Papageorgiou, E.: Supporting meningitis diagnosis amongst infants and children through the use of fuzzy cognitive mapping. BMC Med. Inform. Decis. Mak. **12**(1), 98 (2012)
34. Stach, W., Kurgan, L., Pedrycz, W., Reformat, M.: Genetic learning of fuzzy cognitive maps. Fuzzy Sets Syst. **153**(3), 371–401 (2005)

Chapter 12
Optimising an Agent-Based Model to Explore the Behaviour of Simulated Burglars

Nick Malleson, Linda See, Andrew Evans and Alison Heppenstall

Abstract Agent-based methods are one approach for modelling complex social systems but one issue with these models is the large number of parameters that require estimation. This chapter examines the effect of using a genetic algorithm (GA) for the parameter estimation of an agent-based model (ABM) of burglary. One of the main issues encountered in the implementation was the computation time required to run the algorithm. Nevertheless a set of preliminary results were obtained, which indicated that visibility is the most important parameter in the decision of whether to burgle a house while accessibility was the least important. Such tools may eventually provide the means to gain a greater understanding of the factors that determine criminological behaviour.

1 Introduction

Social systems are incredibly complex due to the large number of interacting elements and many underlying processes that are simply not understood. Moreover, these processes are generally non-linear such that small changes in system parameters can have large effects on the outcomes of the system as a whole. Complex systems are also characterised by self-organisation whereby spontaneous behaviours emerge through the interactions of the individuals and the feedbacks in the system, e.g. the flocking behaviour of birds or the movements of financial markets [19]. Agent-based models (ABMs) have been developed as one technique for modelling complex systems where

N. Malleson (✉) · A. Evans · A. Heppenstall
School of Geography, University of Leeds, Leeds, UK
e-mail: N.Malleson06@leeds.ac.uk

L. See
International Institute of Applied Systems Analysis, Laxenburg, Austria

L. See
Centre for Applied Spatial Analysis, University College London (UCL), London, UK

V. Dabbaghian and V. K. Mago (eds.), *Theories and Simulations of Complex Social Systems*, Intelligent Systems Reference Library 52, DOI: 10.1007/978-3-642-39149-1_12, © Springer-Verlag Berlin Heidelberg 2014

the individuals or 'agents' of the system are explicitly represented in these models. Agents are independent entities that are capable of interacting with each other and with their environment. The agents make assessments of their situation over time (or during each iteration of the model) and then make decisions in response to these assessments [9]. By providing realistic environments and rules that are based on observed and expected patterns of human behaviour, it is possible to create models that can simulate real world systems [69].

Classic examples of ABMs are the Sugarscape model of Epstein and Axtell [32], which simulates wealth accumulation through sugar harvesting in a simple environment, and Schelling's [74] model of segregation, which has been simulated by a number of researchers in the past using an ABM approach (see e.g. [70] and [22]). ABMs are now being applied in a variety of different domains, e.g. ecology [36], economics [78] and more recently, criminology [58–61].

Although ABMs represent a way to capture complexity in social systems, they have issues related to parsimony, i.e. they contain a potentially large number of parameters. Some parameters can be determined through expert knowledge or can be derived from field measurements or social surveys. However, many others are unknown and therefore require a method to determine their values. The need to calibrate a model is not limited to ABMs and many different methods of search and optimisation are available. However, classical search methods are not effective in finding large numbers of parameters so other methods such as genetic algorithms (GAs) are needed.

Despite the fact that GAs are well suited to high dimensional parameter estimation, there are not many examples of the use of GAs in the development of geospatial ABMs. GAs have been used to calibrate cellular automata models of urban land use (e.g. [43, 53, 75]), which might be considered as pre-cursors to ABMs but which are still used today for studying urban form and land use change. One of the most notable examples is the work by Heppenstall et al. [45], who used a GA to calibrate the parameters of an agent-based retail petrol market model. In the model, the market petrol retailers, i.e. the petrol stations, compete for customers within localised, overlapping areas and were therefore represented by agents. Knowledge was embedded in each agent regarding the initial starting price, production costs, and the prices of those stations within their immediate neighbourhood. A series of rules were then applied to each agent in order to effect petrol price adjustments. A GA was used to optimise eight parameters in the model such as the size of the neighbourhood and the production costs. By running the GA several times and examining the variation in the parameters, it was possible to compare which parameters were close in value to those originally determined by the modeller using knowledge of petrol markets and which parameters varied considerably between runs and therefore had little effect on the overall results of the simulations.

More recently, Stonedahl [77] undertook a comprehensive evaluation of GAs for parameter estimation of ABMs in a number of applied areas including archaeology and viral marketing, which showed that GAs can be effective tools for uncovering and further investigating interesting behaviours in these applied areas. However, the research also recommended experimentation with further applications as well as a consideration of multi-objective optimisation problems. The aim of this chapter is to provide an example of single-objective GA parameter estimation in another application area, i.e. crime. An ABM of burglary, which has been previously developed and applied to the city of Leeds in the United Kingdom [59–61], is used to examine the use of a GA for parameter estimation. The chapter begins with an overview of basic crime theory and previous modelling research, including why ABMs are well suited to modelling criminal behaviour. This is followed by a brief overview of optimisation methods and the basic mechanism of a GA. The ABM of burglary is then described including the parameters to be optimised and the GA experimental settings. This is followed by the results of some preliminary experiments and initial reflections upon this method for parameter estimation. The chapter concludes with plans for further research in this area.

2 Theoretical Background

Individual acquisitive crimes are the result of the convergence of a huge number of factors. These include, but are not limited to:

- The motivation of the offender;
- The behaviour of other people including the victim(s);
- The influence of the surrounding physical environment;
- Wider social factors such as levels of community cohesion.

Each of these elements are also extremely complex in their own right. The motivation/behaviour of the offender and other people depends on a wealth of complex psychological characteristics and life experiences as well as factors such as daily routines and transport networks that put people in a particular place at a particular time. The physical environment contains a broad range of 'cues' that might encourage or deter crime (such as high hedges that block visibility, burglar alarms, building security, etc.)—identifying these cues and their impact on offenders is non-trivial. Wider social factors also have a direct influence such as determining how comfortable an offender feels in a particular area (i.e. whether or not they stand out) as well as broader effects that influence where people travel to within a city.

Although the system is clearly complex in the scientific sense of the word, occurrences of crime are not random. Crime patterns can remain stable over long periods of time and a large body of literature has evolved to explain them. This section will outline some of the most relevant criminological findings which form the basis of ABMs of crime as introduced in Sect. 3. As well as demonstrating that the model closely reflects the reality of the real-world crime system, it will make it clear why the ability of agent-based modelling to account for the behaviour and interactions of numerous individuals makes it the most suitable methodology for modelling acquisitive crime and burglary in particular.

2.1 The Spatial Scale of Crime Analysis

Over time, research that seeks to understand the spatial patterns of crime has been moving progressively towards the use of smaller and smaller geographies. In their seminal work on juvenile delinquency, Shaw and McKay [76] used the census tract (an American administrative zone of approximately one square mile). This is roughly the unit of analysis that most modern crime research has continued to use [85], with the exception of some more recent studies that work at smaller census area boundaries of approximately 100–200 households. However, modern environmental criminology theories and recent empirical research [1, 84] suggest that even the smallest areal units of analysis (such as census output areas of less than 1000 people) hide important intra-area crime patterns. As a result of these discoveries, a movement in Environmental Criminology began which focused on the 'micro-places' in which crime occurs [27]. For example, burglars choose individual homes based on their individual characteristics [71] so it cannot normally be assumed that a community or neighbourhood is homogeneous with respect to burglary risk. Similarly, recent work on repeat-victimisation (e.g. [49]) has identified extremely tight spatio-temporal clustering around individual burglary victims. These findings are particularly relevant as most crime modelling research uses aggregate data that hide these important micro-level patterns (see Sect. 2.3 for more details).

2.2 Environmental Criminology Theories

The movement in crime research towards using individual-level geographies also resonates with the major theories in Environmental Criminology. As this section will illustrate, these theories focus specifically on the spatio-temporal behaviour of the individual(s) involved in crime events and the intricacies of the immediate surrounding physical environment.

Routine activity theory [21] explores the interactions between victims, offenders and other people who might influence an individual crime event (e.g. passers-by, police, etc.). For the crime to occur, the theory stipulates that an offender must meet a victim at a time and place with an absence of others who might prevent the crime. This convergence depends on the routine activities of the people involved. For example, a burglar might come into contact with a vulnerable house (the potential victim), but might not be able to commission a crime if the routine activities of the residents or neighbours mean that they are in the area at the same time and will notice a crime taking place.

The geometric theory of crime [10] shares many similarities with routine activities theory, but focuses more explicitly on the interdependencies between a person's knowledge of the environment, i.e. their *awareness space*, and criminal opportunities. The theory considers how the routes used to travel around a city influence a person's awareness space and hence the spatio-temporal locations in which offenders are likely to commit a crime. Burglars do not search for targets at random; instead they are likely to search near important 'nodes' such as friends' houses, schools, work places, or places of leisure [11]. Thus house vulnerability to burglary is less relevant if the house itself is not within the awareness space of a person who might attempt to burgle it.

The final theory that the model logic attempts to replicate is the rational choice perspective [20]. This suggests that the offender's decision to offend is a cost-benefit analysis weighing up potential rewards of a successful crime with the risks of being apprehended. Thus a crime will only be committed if it is perceived as profitable. It is important to view the concept of rationality as 'bounded', such that a decision that might appear to be optimal to one person (in a specific situation with their own thoughts and motivations) might be blindingly irrational to another.

Although they describe different elements of the crime system, the theories largely agree on the mechanisms that lead to the spatio-temporal patterns of crime.It is particularly relevant that each theory emphasises the *individual-level nature* of crime occurrences. The crime system is driven by the behaviour and interactions of individual people situated in a highly detailed local environment. Aggregating such a system (either spatially or temporally) will hide important lower-level dynamics that ultimately explain why crime takes place in the places that it does.

2.3 Traditional Crime Models

Traditionally, quantitative crime models have used area-based crime data in regression style modelling (see e.g. [12]). Kongmuang [52] provides a comprehensive review of the methods employed, from which a number of common characteristics can be identified. For example, model accuracy is usually estimated through the Akaike Information Criterion (AIC) or a goodness-of-fit statistic such as R^2. Some drawbacks are outlined below although we do recognise that there are also many advantages of statistical methods which are not discussed further in this chapter.

Firstly, statistical models generally utilise simple functional relationships, e.g. they cannot adequately capture the evolution of individuals through time and the effect this has on their behaviour. In contrast, ABMs can represent these complex real world interactions including the intricate personal trajectories and histories of individuals. Statistical techniques generally aim to reduce variables to enhance explanation at a cost to predictive power, so cannot account for the complexity of the environmental backcloth and the non-linear human-human or human-environment interactions that drive the system.

Secondly, the use of spatially aggregated data—to represent crimes, demographics, the environment, etc.—hides important lower-level relationships between crime, individuals and the environment. Similarly, it is difficult to capture important features of the physical environment such as accurate travel times, impassable barriers or road-network layout unless individual environment objects (roads, buildings, parks, etc.) are accounted for explicitly.

Finally, linear models may be "computationally convenient" [28], but they cannot represent the dynamics of complex systems. Complex systems are driven by the behaviour of and interactions between the individual components of the system. These fundamental drivers of the system are lost when the underlying data are aggregated.

In general, the dynamics that drive the *crime system* (as with other social systems) are not captured directly in aggregate models. This makes it difficult both to explore criminology theory—which inherently focuses on the spatio-temporal behaviour of individual people—and to make crime forecasts at the same time. ABMs, however, provide an alternative approach by allowing these individual entities to be modelled *directly*. In this manner, it is possible to capture the true richness of the system and much more closely reflect an individual's unique circumstances and behavioural characteristics.

3 Agent-Based Models of Crime

ABMs have a number of clear advantages over other modelling and analysis techniques when it comes to understanding crime. Crime tends to be the result of individuals acting on the basis of their history and current environment, either alone or in collaboration. ABMs, unlike other techniques, take as their starting point unique individuals ('agents') with their own history and decision making capacities, and these individuals are placed in a complicated environment to discover the resultant behaviour. As in a real crime system, agents will both respond to and adjust the current environment (e.g. agents may cause an area's attractiveness to housebuyers to fall). The agents in ABMs can interact and collaborate in group behaviour and decision making. However, there is also no reason why larger, aggregate groupings and decision making (e.g. government policy groups) cannot also be represented and respond to the system. In short, ABMs represent social systems in the way we intuitively understand social systems ourselves.

This does not, of course, *necessarily* make such models better ways of understanding such systems. However, in practice there are considerable advantages to matching our understanding of reality as closely as possible. Firstly, ABMs allow for the direct representation of decision making using rulesets that act at the individual level. This means that the errors associated with representing behaviour are less likely than, for example, if such behaviours were represented as aggregate mathematics. The concentration on rulesets and behaviour also means that ABMs can act as a framework for the representation and testing of qualitative social theory described at the individual level, something much harder to achieve with mathematical or statistical representations. ABMs act as a framework for understanding emergence, that is, how behaviour at the individual level can generate complex patterns at some larger scale (like crime hotspots). Secondly, agents can have an individual history. Statistical techniques are limited in their ability to track how life-events and the environment interact. With ABM, individuals carry their history with them, either implicitly or explicitly, and this history can be analysed to see how it affects their decision making. Finally, ABMs can represent a wide range of environments, from the very abstract, to the extremely realistic. This allows us to explore and understand the effects of the environment on behaviour at a very detailed level. For example, it is possible to look at the effect that a specific change in a public transport route might have on criminal opportunity. Moreover, once an ABM is set up, a wide variety of different analyses and scenarios can be run without adjusting the underlying model, unlike many other techniques, where the model must be specifically designed from the ground up to answer a single research question.

Given these advantages, it is somewhat surprising how slowly the development of ABM of crime has progressed. Nevertheless, the last ten years or so has seen an increased interest in the technique, and a number of groups are building ABMs of crime at various levels of detail. In general, most current ABMs of crime attempt to replicate the major components of the criminal system to some degree: offender motivation and decision-making, offender behaviour and movements, victimhood and guardianship. However, given the complexity of the system, it should come as no surprise that most concentrate on building realism in one of these components rather than all of them. In addition, the realism of the environment within the model varies a great deal, not least because of the broad division in agent-based modellers between those who believe ABM should be utilised as abstract 'thought experiments' to explore key theoretical behaviours and ideas, and those who believe that it is possible to build a more detailed model of the real world for exploration and prediction (see, for example, Di Paolo et al. [25], for arguments for the former).

Malleson et al. [62] give a full review of ABMs used to model crimes that have a predictable geographical component (that is, crimes like burglary and street theft, as opposed to crimes like domestic violence and fraud, on which geography have less obvious effects). However, notable models at the more abstract end of the scale include Winoto [87]; van Baal [80]; Brantingham and Brantingham [13]; Brantingham et al. [14–16]; Dray et al. [26], and Wang et al. [83], while more realistic models have been attempted by Liu et al. [54], Melo et al. [65], Birks [5–8], Malleson et al. [56, 58]; Groff [38–40], Groff and Mazerolle [41], and Malleson [57], Malleson et al. [59–61] and on social, but not geographical realism, Hayslett-McCall et al. [44]. In addition, the technique is seeing a growing use in modelling crimes where geography is secondary to social organisation, e.g. gang crime and civil violence [4, 18, 29, 48, 55]. Some of the ethical issues facing agent-based modellers of crime are explored by Evans [33].

4 Optimisation of Complex Models

We now consider the specific issue of parameter estimation in ABMs using methods of optimisation. There are techniques that search a problem space for the best solution possible given the complexity of the problem, the computational resources available and the objectives or constraints of the problem [42]. Mathematically this involves finding what are referred to as a set of decision variables that minimise or maximise one or more objective functions (i.e. a function specific to the problem which determines how good the solution is) subject to satisfying a set of constraints. For example, the decision variables may be the quantity of material that flows between a set of different distribution points where the objectives are to minimise the distance travelled while maximising the profit subject to certain routes not being allowed due to direction of flow or excessive gradients. Some problems may have a single optimal solution where the main challenge is finding the global optimum in a solution space characterised by multiple local minima (or maxima depending upon the way the problem is formulated) without having to fully search the whole parameter space. In contrast, in more complex problems or in those with multiple objectives, there is no single solution that simultaneously optimises all conflicting objectives. The result is a set of alternative optimal or feasible solutions of similar fitness that represent trade-offs between the different objectives. Optimisation provides the mechanism to find this set of solutions, which are called Pareto optimal solutions. Other methods, such as multi-criteria decision making, are then needed to further evaluate the solutions that are identified during the optimisation process. Many real world problems are characterised by the need to take conflicting multiple objectives into account. In hydrology, for example, multi-objective optimisation methods are used extensively for calibrating physical and conceptual hydrological models [31, 82, 89].

Optimisation can be divided into the following seven steps: identify the parameters in the problem or model; choose the design variables (or those which require optimisation) from this set of parameters; outline any constraints which must be taken into account during the optimisation; choose appropriate objective functions, i.e. methods of evaluating the solution or model performance; set the allowable range for the decision variables; choose an appropriate optimisation algorithm; and run the algorithm to obtain the results [24]. The next section deals specifically with step 6, i.e. different methods of optimisation.

4.1 Methods of Optimisation

A number of different optimisation methods have been developed in the past to handle problems involving single and multiple objectives. Classical (or conventional) optimisation methods were developed using differential calculus. They involve finding an analytical solution on functions that are continuous and differentiable, e.g. the simplex method [64]. These are referred to as strong methods and are deterministic. On the other end of the spectrum are weak classical methods, which involve a random or stratified random sampling of the search space in order to find the solution, which are inefficient methods. These classical methods have a number of disadvantages [24, 42]. For example, the convergence to an optimal solution depends upon the initial solution and they are not efficient for problems with discrete rather than continuous search spaces. Moreover, they are not efficient in solving non-linear, complex problems with large search spaces and many conflicting objectives and they cannot be parallelized efficiently since they use a single search path to obtain the optimal solution. For this reason, a set of intermediate methods have been developed that contain a stochastic element and which use more effective search strategies to avoid being trapped in local minima, e.g. simulated annealing, tabu search and evolutionary methods such as genetic algorithms (GAs). The focus of this research is on GAs, which are described in more detail in the sections that follow.

4.2 Genetic Algorithms

GAs are intrinsically suited to optimisation when the fitness landscape is complex, changes over time or has many local optima. Through inherent parallelism, they are able to simultaneously explore numerous potential solutions [42, 47, 68]. Within in GA, unique combinations of parameter values are represented as binary strings. These individual strings are also referred to as 'chromosomes' and comprise a combination of 'genes' (the individual parameter values). Through the evaluation of the fitness of one string, a GA is also simultaneously sampling each of the many other spaces to which it belongs. Over several fitness evaluations, the GA builds up an increasingly accurate value of the average fitness of each of these spaces. Through the evaluation of a small number of individuals, a much larger group is being evaluated implicitly. By this mechanism, a GA can 'home in' on the space with the highest-fitness individuals. This combination of parallelism, along with the other major components of a GA which produce the evolution of fitter solutions, i.e. selection, mutation and crossover, make this approach a very powerful and efficient tool.

GAs follow the same basic set of steps as outlined in Fig. 1. A population is first initialised and the objective functions are then set. The fitness of each individual is assessed and on the basis of this, the fittest in the population are selected for reproduction via crossover. This continues over many generations or iterations until predefined criteria are satisfied, e.g. a certain threshold value for the objective function has been reached. For a more detailed overviews of GAs, the reader is referred to Goldberg [42], Davis [23], Michalewicz [66], Bäck and Schwefel [2] and Eiben and Smith [30]. For an overview of GAs in the context of geographical optimisation, Xiao [88] provides an excellent introduction.

The following sections will briefly outline the main generic parameters and processes that all GAs share.

4.2.1 Initial Population

At the start of an optimization, the GA requires a set of initial solutions. There are two ways of forming this initial population. The first involves randomly generating solutions while the second uses some expert knowledge about the problem. The advantage of the second method is that the GA starts with a set of approximately known solutions and therefore may converge to an optimal solution faster than the

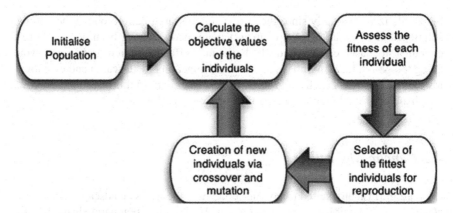

Fig. 1 The basic operation of a GA (adapted from [86])

first method. The disadvantage is that genetic diversity may be restricted and limit the ability of the GA to generate optimal solutions that might only be arrived at through a random starting position.

4.2.2 Representation

Most of the problems suitable for GAs involve identification of a set of parameters that need to be represented in such a way as to allow evolutionary operators to be effectively applied. As GAs are robust, there is little need to rigorously identify the 'best' representation for a particular problem [42]. There are two broad methods that can be used for representation: binary alphabets [46] and real numbers [3, 23, 66, 67]. There is no single 'correct' coding method for encoding a problem; the mode of representation is dependent on the problem. However, the coding sequence must adequately represent the problem to ensure that the optimal solution is available to the algorithm and be bounded by an allowable range for the parameters.

4.2.3 Fitness and Selection

In order to evolve better performing solutions, the fittest members of the population are selected and randomly exposed to mutation and recombination (as described below). This produces offspring for the next generation. The least fit solutions die out through natural selection as they are replaced by new recombined, fitter, individuals. Evaluation of the fitness of the individuals involves some form of comparison between observed and model data, or a test to see if a particular solution meets pre-defined criteria or constraints. In this work, the Standardised Root Mean Square Error (SRMSE—[51]) is used to estimate the difference between real crime data and the model results.

Table 1 Description of several of the most common forms of parental selection

Selection type	Description
Ranking	The population is sorted from best to worst. The number of copies that an individual receives is given by an assignment function and is proportional to the rank assignment of an individual
Tournament	A random number of individuals are selected from the population. The best individual from this group is chosen as a parent for the next generation. This process is repeated until the mating pool is filled
Roulette wheel	Individuals are mapped to contiguous segments of a line, such that each individual's segment is equal in size to its fitness. A random number is generated and the individual whose segment spans the random number is selected. This process is repeated until the desired number of individuals is obtained
Truncation	Truncation sorts individuals according to their fitness (from best to worst). Only the best individuals are selected to be parents

There are number of possible ways for selection to take place and Table 1 describes the main parental selection schemes that recur within the literature.

4.2.4 Selection Pressure

Along with the selection method, the selective pressure parameter is critical. This parameter measures the probability of the best individual being selected compared to the average probability of selection and drives the algorithm towards a solution. The value for this parameter should be carefully selected as too much selective pressure can lower the diversity within the population resulting in sub-optimal solutions. Conversely if the selection pressure is too low, the population remains too diverse and the optimal solution is not found.

4.2.5 Recombination/Crossover

The main reproductive genetic operator is recombination (also known as crossover). This is the process by which new individuals are produced by combining the information from two parent chromosomes. The resulting offspring inherits components from both parents (Fig. 2). This allows the EA to explore new areas in the search space. Without recombination, the offspring are simply duplicates of the parents, which does not provide the opportunity to improve the fitness of the population.

There are several methods of recombination available; the suitability of the method is dependent on the types of genes or variables stored in the chromosome. The three most common approaches are intermediate, line and extended line recombination methods. In intermediate recombination, the variable values of the offspring are randomly chosen from between the values of the parents. Normally values of up to 25 %

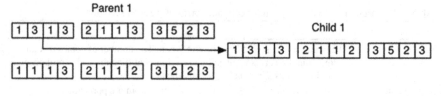

Fig. 2 Representation of recombination between two parents to produce an offspring

Fig. 3 Illustrating the mechanism of mutation

outside this range can be used, which has been chosen to ensure that statistically a space covered by the recombination does not decrease in size with time leading to a loss in diversity. The position of the variable chosen on the line determines how much each parent contributes to the offspring and is chosen uniformly at random for each gene. Line recombination is similar to intermediate recombination except that the same random number is used for selecting the value of every gene in a chromosome. Extended line recombination is different from the above techniques in that the variable range is not limited to a range around the parents. The probability of any particular value being taken is not uniform but varies with a high probability near the parents and a low probability far away from the parents. The probability distribution can also be chosen to favour the fitter parent. The value controlling the amount of the parent that is used is generated randomly and then used for selecting the value of subsequent genes.

4.2.6 Mutation

The process of recombination can produce a very large number of new individuals. However, if the GA is moving towards an optimal solution (and hence a smaller population pool), it is possible that the available solutions are suboptimal. Through the alteration of one or more parts of the chromosome, mutation introduces diversity into the selected population which can potentially breed fitter solutions (Fig. 3). The mutation rate is generally a random probability determined by initial experimentation.

The literature offers no strict guidelines for the selection of the size of the mutation step. The optimal step-size depends on the research problem and may even vary during the optimisation process. Small mutation steps are acknowledged in the literature as being successful, especially when the individual is already well adapted. However, large mutation steps can, when successful, produce good results very quickly. A good mutation operator should therefore produce small step-sizes with a high probability and large step-sizes with a low probability.

In the next section, the ABM burglary model is introduced along with the settings of the GA for parameter estimation.

5 The Agent-Based Burglary Model

The model utilised here attempts to provide a detailed burglary model at the city scale that includes (a) detailed offender drivers, decision making, and behaviour; (b) realistic victim distributions and attributes, including daily variations in household occupancy; and (c) a realistic environment including a full transport network and reasonable levels of guardianship, including community guardianship. A full ODD protocol [36, 37] description of the model can be found in Malleson et al. [63], while a detailed description of the model design and data preparation is given in Malleson [57]. The full model uses the PECS framework [72, 73, 79] for internal offender decision making can be found in Malleson et al. [60]. However, to simplify the model for the application of the GA, a simpler behavioural framework was implemented, the details of which are outlined below. Calibration and validation of the model were carried out manually in the past; details can be found in Malleson et al. [61]. Here, we briefly describe the model in sufficient detail (using a simplification of the ODD protocol) to understand the broad model workings and how the parameters that will be calibrated fit into the model.

5.1 Purpose of the Model

The motivation behind the model is to simulate the spatio-temporal locations of burglaries at the city scale and, ultimately, to provide a framework for modelling and testing our understanding of the criminal system. The model runs for a fixed length of simulated time—sufficient to reach dynamic equilibrium—so does not predict the actual number of crimes. Instead, we focus here on the values of the behavioural parameters that drive the behaviour of the agents to determine what these tell us about the behaviour of burglars in the real world.

There is no notion of the processes that lead to someone 'becoming' a burglar; each agent has only one purpose which is to commit burglary. In addition, there is no notion of punishment or capture—offenders are not removed from the system, nor are their drivers adjusted by any kind of punishment. Although variables such as community guardianship help to determine whether a property is chosen for a crime, a chosen target is always successfully victimised. There is also no communication between agents; all offenders are currently lone individuals without a shared understanding.

The model generates a spatial distribution of crimes, taking into account a variety of offender behaviours, environmental factors, and victim and guardian attributes.

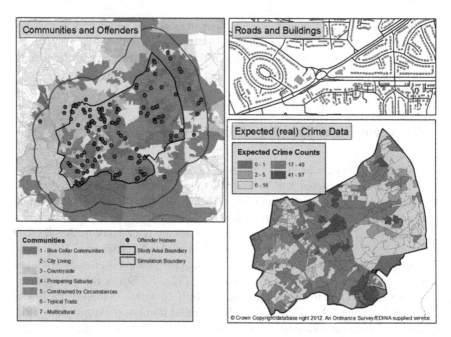

Fig. 4 Datasets used in the ABM of burglary

5.2 Data and the Study Area

The study area for this research covers 1,700 hectares in the city of Leeds, UK. The area contains some of the most deprived neighbourhoods in the country and was earmarked for an ambitious urban renewal scheme which makes it an ideal candidate for predictive crime modelling. Figure 4 illustrates the data used to represent the study area in the simulation.

- *Communities* are generated using the Output Area geography (a census area boundary containing approximately 100 houses) and classified using the Output Area Classification (OAC: [81]). This allows community types to be compared quantitatively.
- The home locations of *offenders* were estimated from police recorded crime data on convicted burglars.
- *Roads* and *buildings* are established from Ordnance Survey MasterMap data (the Integrated Transport Network and Topographic Area data sets respectively)
- *Expected crime data* are required to validate the model. The data used here are the number of burglaries per Output Area that occurred in 2001. This year was chosen because it corresponds closely with the timing of the UK census from which community demographics are estimated. As mentioned in Sect. 4.2.3, the SRMSE is used to compare model results to expected data at the output area level.

Finally, a buffer zone is used to reduce the boundary effects in the results. Thus all burglaries that occur outside the study area boundary are discounted when simulated burglaries are compared to real data.

5.3 State Variables and Scales

5.3.1 The Agents

The model is comprised of agents representing offenders. Victims and guardians are represented through environmental variables, e.g. the estimated level of community cohesion. Each offender agent is assigned a home building (and associated community) at model initialisation. This location, derived from real offender data, is where the agent lives. The main agent variable, which changes during the model run, is the *burglary motive*. This variable increases over time and determines whether or not a burglar will choose to target an individual house; more details follow in Sect. 5.3.3. Once a burglary has been committed, this level falls to zero.

5.3.2 The Environment

Objects within the environment build up a substrate in which the agents act. There are three types of objects:

- **Roads** are used to restrict the possible spatial locations of the agents. In the full version of the model, roads can be used to simulate different transport speeds and routes (e.g. a car driver moving faster along a major road) but in this simplified version all agents move at a constant speed of 4 miles per hour (a fast walking pace).
- **Buildings** represent the houses in which agents live and also represent the potential victims of burglary. Their spatial locations and attributes have been established from the Ordnance Survey MasterMap geographic data product.
- **Communities** represent the neighbourhoods in which the houses are located. They will influence whether or not a burglar chooses a particular house and also determine where an offender starts their search.

Section 5.2 outlined the data sources that have been used to generate these layers while Table 2 summarises the fixed attributes for the buildings and communities. As Sect. 5.3.3 will illustrate, these will influence an agent's burglary decision about where to look for victims and which individual building to target.

5.3.3 Process Overview and Scheduling

The model is initialised with data that allocates offenders to households, attributes to buildings and transport components, and initialises the state variables of the offenders. Offenders start with nothing in their awareness space. At each time step, all

Table 2 Parameters associated with communities and buildings

Parameter (abbreviation)	Description
Communities	
Attractiveness (ATT)	A measure of the abundance of valuable goods that is likely to be found in houses within the community. This measure was calculated from factors such as the percentage of full time students and the percentage of houses with more than two cars
Social type (SocT)	A vector containing the 41 different OAC variables. This can be used to compare the Euclidean difference between two communities
Collective efficacy (CE)	A measure of the cohesion of the community, calculated from a combination of deprivation, residential stability and ethnic heterogeneity
Buildings	
Accessibility (ACC)	An estimate of how easy a building is to enter based on its spatial properties. Houses with many exposed walls are assumed to contain a larger number of doors or windows to provide access to a burglar
Visibility (VIS)	A measure of how visible a building is to its neighbours and to passers-by. This is calculated from the size of any attached garden and the number of additional buildings within a buffer zone

offenders decide on actions determined by their internal states. The sequence of offenders is random.

In its full implementation, the model uses an advanced behavioural framework to equip the agents with realistic daily behaviour including sleeping, using substances and socialising. However, the focus of this work is to better understand the relationship between the behavioural parameters and simulated burglary locations so the behaviour of the agents has been simplified by removing non-burglary activities. Therefore, activities such as socialising or using substances will not have an influence on the final locations of the burglaries. As mentioned previously, each agent is driven by a single 'burglary motive', which increases over time until a burglary is committed. A burglar's behaviour schedule is as follows:

1. At 09:00 simulated time, the agent chooses a community to travel to in search of a burglary target. The following formula is used to assign the likelihood, L, of choosing each community, a, relative to their current location, c, and their home community, h:

$$L = w1^*(1/\text{dist}(c, a)) + w2^*\text{Attract}(h, a) + w3^*\text{SocialDiff}(h, a)$$
$$+ w4^*\text{PrevSucc}(a) \tag{1}$$

where dist(c,a) represents the distance (travel time) from their current location to the target community, Attract(h,a) represents the relative attractiveness of the community compared to the agent's home area, SocialDiff(h,a) returns the similarity of the target community and the agent's home (where similar communities are favourable) and PrevSucc(a) returns the number of times an agent has had a successful burglary in the past for a particular community, a. Importantly, the weights $w1$ to $w4$ can be used to assign an importance to each factor—if a weight is large then the parameter will have a greater influence on the agent's behaviour. Roulette-wheel-selection is used to randomly choose a community from all of those available such that those with a greater likelihood value have a greater chance of being chosen.

2. On the way to their destination, the agent observes each house that they pass and a 'risk' for burglary is calculated as follows:

$$R = (w5^*CE + w6^*ACC + w7*VIS)/(w5 + w6 + w7) \qquad (2)$$

where CE is the perceived collective efficacy of the surrounding area, ACC represents the accessibility of the target building (how easy it is to enter) and VIS represents the visibility of the building to neighbours and passers-by (where high visibility increases the risks). If this risk value is lower than the agent's current burglary motive, then the agent *might* commit a burglary—the probability of actually committing a burglary increases exponentially as the difference between risk and motive increases. Again, weights applied to each parameter determine how much of an influence each factor will have over the agent's decision.

If the burglary is successful, then the agent travels home. In this manner the model has been configured to allow for a variation in offending behaviour (the same agent will not always choose the same house) but agents will, on average, always commit one burglary per day.

3. If the agent reaches their chosen destination without committing a burglary, then they repeat the process. This continues until a victim has been found or another need becomes greater than the need to burgle, such as the need to sleep, which results in the burglar travelling home. Thus there may be days when no burglaries are committed.

6 Exploring the Dynamics of Criminal Behaviour

The model outlined in Sect. 5 is an advanced ABM that attempts to closely represent criminological theory and the experiences of crime-reduction experts in the field. There are 7 different variables that determine where burglar agents will start searching for targets and which houses, in particular, they will actually victimise. Although some experimentation with changing the parameters can be undertaken manually through trial-and-error, the number of combinations that can be tested, even with 7 parameters, is limited. Thus, the GA provides a more comprehensive approach to more intelligently explore the entire 7-dimension parameter space. This can help to determine which parameters have the most substantial influence on the model, and the values may eventually inform our understanding of the behaviour of burglars in the real world.

A set of preliminary experiments have been undertaken here using a GA to find the values of the 7 parameters. A population of 20 chromosomes was used and the GA was tracked over three iterations or generations. The reason for 3 generations and 20 individuals (or 20 model runs) was dictated by computational need. One major issue with the GA approach as presented in this study is the large amount of computational time required to run the model, even with only three iterations per GA run. First, the ABM itself is extremely computationally expensive. Even after some simplifications from the original configuration [60], a single model run still required approximately 10 h to complete on a normal desktop machine. The results discussed here were generated with the use of a 16-core Intel Xeon E5-2670 ("Sandy Bridge") virtual machine provided through Amazon Web Services [34], but even with this hardware, each GA iteration—with a population of only 20 chromosomes—required approximately 20 h to complete or more than 60 h to run a single experiment. As a result, it was not feasible to run each individual model configuration multiple times, which would be preferable because it would give a more comprehensive assessment of the model error (since the simulation is probabilistic and each run will therefore lead to slightly different results).

The fitness of each model configuration (or 'chromosome') is provided in Fig. 5, which is plotted against the iteration or generation number. As would be expected, the GA is able to identify which model configurations result in the lowest error and, hence, which should be used to generate the configurations in the next iteration or generation. Accordingly the model error, which is the SRMSE between the real crime data and the model results, decreases with each subsequent iteration. Figure 5 also highlights some clustering of fitness values after the initial (random) population undergoes an evolution. This illustrates that the algorithm is fine-tuning the ABM in different parts of the parameter space that have the lowest error.

Table 3 provides the values for each of the parameters for the models with the lowest error after each iteration. Due to stochastic elements in the model, model configurations with the same weight values can result in slightly different values for the fitness. The GA appears to converge very quickly to an optimal configuration, which is found after the first iteration and does not change substantially over the next few iterations, although a marginally different configuration is found in iteration 3, which may have been the result of the GA exploring a slightly different part of the solution space. This implies that the algorithm has found a global maximum. Alternatively, the model may simply need to run many more generations, but as discussed previously, the computational load is a real issue in using a GA for parameter estimation of this particular burglary ABM.

Figure 6 illustrates the change in parameter value for these best models. Interestingly, the *visibility* parameter is consistently assigned a high value, which implies it is an important factor in the agent's decision compared to other variables. This means that with the models that closely matched the observed crime data, the agents were more likely to burgle houses that were well hidden from their neighbours and passersby. In the ABM used here, this feature was estimated by combining the phys-

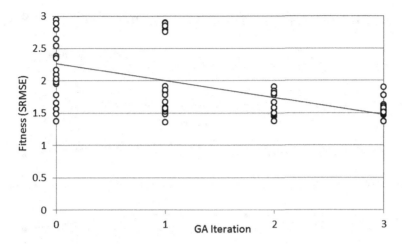

Fig. 5 Fitness of all the chromosomes by GA iteration

Table 3 Values of the parameters after each iteration

Iteration	Fitness	w1	w2	w3	w4	w5	w6	w7
0	1.372	0.719	0.668	0.736	0.683	0.541	0.291	0.984
1	1.362	0.719	0.668	0.736	0.683	0.541	0.291	0.984
2	1.372	0.719	0.668	0.736	0.683	0.541	0.291	0.984
3	1.365	0.689	0.717	0.781	0.727	0.510	0.241	1.000

The weights: w1 = distance; w2 = attractiveness; w3 = SocialDifference; w4 = PreviousSuccess; w5 = CollectiveEfficacy; w6 = Accessibility; w7 = Visibility

ical size of a house's garden with a measure of its isolation (i.e. the number of other properties in the immediate surrounding area). Conversely, the parameter accessibility appears to have only marginal importance and was consistently the least important of the seven parameters. This parameter is calculated by estimating the number of exposed walls in each building such that detached houses are at a greater risk of being burgled than semi-detached houses or terraces because there are likely to be a larger number of entry points. *Why* this building feature appears to be less important, however, is less clear. The parameter *collective efficacy*, which is a measure of the cohesion of the community and is calculated from a combination of deprivation, residential stability and ethnic heterogeneity, is also of less importance than visibility but of greater importance than accessibility, where these three parameters together are used to calculate the risk associated with choosing a property to burgle. Thus it appears that once a community has been chosen, the primary driving force is the visibility of the property. The other four parameters all had similar weights, i.e. on the order of 0.68 to 0.72, where each of these parameters, i.e. attractiveness, social type, previous success and distance to the target location, are all used to calculate the likelihood of choosing a community. This suggests that these factors are equally important in making a decision about a particular area to target. The inclusion of

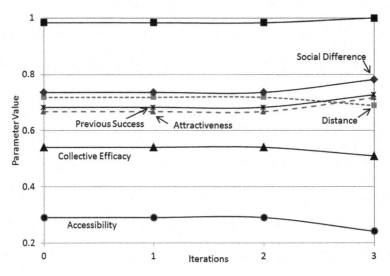

Fig. 6 Parameter values for the best model after each GA iteration

these seven parameters is based on criminology theory but there is little information about the importance of these parameters, which is one area where optimisation with GAs may shed some evidence.

To review the results spatially, Fig. 7 presents maps of the burglary counts generated by the three best model configurations after GA iterations one ('Model 1'), two ('Model 2') and three ('Model 3'). The results are presented in two forms. The maps on the left hand side of Fig. 7 present results that have been spatially aggregated to the geography of the communities (i.e. each community has a count of burglaries that were committed during the model run) and the maps on the right hand side of Fig. 7 present point density estimates produced using the Kernel Density Algorithm (KDE). The use of KDE arguably presents a more accurate picture of the underlying point patterns and is commonly used by police analysts [17].

The model results show consistently high numbers of burglaries on the western side of the study area, which matches the general pattern exhibited by the observed data. Interestingly, these larger scale patterns are similar regardless of the differences in configuration, which suggests that small changes to any of the agents' behavioural parameters have little effect on the model results. Some small differences can be seen in the centre of the study area where some small hotspots are picked up differentially between the three models. Any discrepancies are likely to be a result of the probabilistic nature of the model (recall that with sufficient computing power each model would have been run a large number of times to calculate an average error value) although there would be scope to investigate what might be generating these differences, e.g. the slight decrease in the distance, collective efficacy and accessibility parameters in 'Model 3' and a slight increase in all the others. However, it is encouraging that small parameter changes have little effect on the model results because,

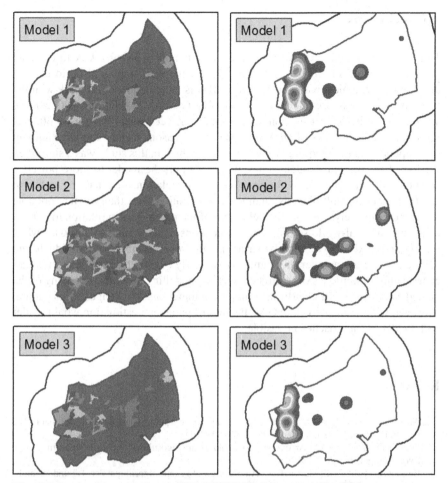

This work is based on data provided through EDINA UKBORDERS with the support of the ESRC
and JISC and uses boundary material which is copyright of the Crown.

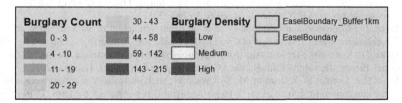

Fig. 7   Results from the three best model iterations after 1, 2 and 3 iterations

were this not the case, it would be more difficult to be confident in the robustness of
the results. This represents another considerable advantage of the application of an
optimisation algorithm to this model.

7 Conclusions

This paper presented some very preliminary attempts at using a GA to estimate the parameters of an ABM of burglary. One of the next steps will be to explore the seven geographical parameters of the ABM is more detail and to work with the full behavioural-based model in the future. The GA provides the means to explore a very large solution space in an efficient manner, yet has not been used often to determine the parameters of an ABM. One of the reasons may be the large amount of computational time required to run an ABM, which in this study was a simplified version of the original model. Future work will port the model to more powerful computer systems and run a GA with a larger population for a larger number of iterations and with multiple runs per model configuration. Only then will it be possible to more exhaustively examine the behaviour of the parameters in relation to burglar behaviour in a real-world setting. The implications of using such an approach when scaling up an ABM to a much larger area, with larger numbers of individuals and with a much larger number of parameters is clearly evident from these preliminary experiments. Ambitious ABM projects such as modelling the entire economy of the United States [35] will clearly face major computational challenges in the future. However, without methods like GAs, the task of parameter estimation would render such modelling approaches infeasible.

References

1. Andresen, M., Malleson, N.: Testing the stability of crime patterns: implications for theory and policy. J. Res. Crime Delinquency **48**(1), 58–82 (2011)
2. Bäck, T., Schwefel, H.-P.: An overview of evolutionary algorithms for parameter optimization. Evol. Comput. **1**(1), 1–23 (1993)
3. Beasley, D., Bull, D.R., Martin, R.R.: An overview of genetic algortihms: Part 1. Fundamentals. Univ. Comput. **15**(2), 58–69 (1993)
4. Bhavnani, R., Miodownik, D., Nart, J.: REsCape: an agent-based framework for modeling resources, ethnicity, and conflict. J. Artif. Soc. Soc. Simul. **11**(2), 7 (2008) [Online]. http:// jasss.soc.surrey.ac.uk/11/2/7.html. Accessed 16 June 2011
5. Birks, D.: Computational criminology: a multi-agent simulation of volume crime activity. Presentation to the British Society of Criminology Conference, University of Leeds, UK (2005)
6. Birks, D.: Synthesis over analysis: using agent-based models to examine the interactions of crime. Presentation at the Fifth National Crime Mapping Conference, London (2007)
7. Birks, D.J., Donkin, S., Wellsmith, M.: Synthesis over analysis: towards an ontology for volume crime simulation. In: Liu, L., Eck, J. (eds.) Artificial Crime Analysis Systems: Using Computer Simulations and Geographic Information Systems, pp. 160–192. Information Science Reference, Hershey (2008)
8. Birks, D., Townsley, M., Stewart, A.: Generative explanations of crime: using simulation to test criminological theory. Criminology **50**(1), 221–254 (2012)
9. Bonabeau, E.: Agent-based modeling: methods and techniques for simulating human systems. Proc. Nat. Acad. Sci. **99**, 7280–7287 (2002)
10. Brantingham, P.J., Brantingham, P.L.: Notes on the geometry of crime. In: Brantingham, P.J., Brantingham, P.L. (eds.) Environmental Criminology (pp. 27–54). Waveland Press, Prospect Heights (1981)

11. Brantingham, P.L., Brantingham, P.J.: Nodes, paths and edges: considerations on the complexity of crime and the physical environment. J. Environm. Psychol. **13**(1), 3–28 (1993)
12. Brantingham, P.L., Brantingham, P.J.: Mapping crime for analytic purposes: location quotients, counts, and rates. In: Weisburd, D., McEwen, T. (eds.) Crime Mapping and Crime Prevention, Volume 8 of Crime Prevention Studies, pp. 263–288. Criminal Justice Press, Monsey (1998)
13. Brantingham, P.L., Brantingham, P.J.: Computer simulation as a tool for environmental criminologists. Secur. J. **17**(1), 21–30 (2004)
14. Brantingham, P.L., Glasser, U., Kinney, B., Singh, K., Vajihollahi, M.: A computational model for simulating spatial aspects of crime in urban environments. Proceedings of the 2005 IEEE International Conference on Systems Man and Cybernetics **4**, 3667–3674 (2005)
15. Brantingham, P.L., Glasser, U., Kinney, B., Singh, K., Vajihollahi, M.: Modeling urban crime patterns: viewing multi-agent systems as abstract state machines. Proceedings of the 12th International Workshop on Abstract State Machines, pp. 101–117, Paris (2005)
16. Brantingham, P.L., Glasser, U., Jackson, P., Kinney, B., Vajihollahi, M.: Mastermind: computational modeling and simulation of spatiotemporal aspects of crime in urban environments. In: Liu, L., Eck, J., (eds.) Artificial Crime Analysis Systems: Using Computer Simulations and Geographic Information Systems, chapter 13, pp. 252–280. IGI Global, Hershey (2008)
17. Chainey, S., Ratcliffe, J.: GIS and Crime Mapping, 1st edn. Wiley, Chichester (2005)
18. Cherif, A., Yoshioka, H., Ni, W., Bose, P.: Terrorism: Mechanisms of Radicalization Processes, Control of Contagion and Counter-Terrorist Measures. Santa Fe Institute Working Paper [Online] (2009). http://tuvalu.santafe.edu/events/workshops/images/7/7e/TerrorismWorkingPaper.pdf. Accessed 16 June 2011
19. Cilliers, P.: Complexity and Postmodernism. Routledge, Bury St Edmonds (1998)
20. Clarke, R.V., Cornish, D.B.: Modeling offenders' decisions: a framework for research and policy. Crime Justice **6**, 147–185 (1985)
21. Cohen, L., Felson, M.: Social change and crime rate trends: a routine activity approach. Am. Sociol. Rev. **44**, 588–608 (1979)
22. Crooks, A.: Constructing and implementing an agent-based model of residential segregation through vector GIS. CASA Working Paper Series, Paper 133 (2008). http://eprints.ucl.ac.uk/15185/1/15185.pdf
23. Davis, L.: Handbook of Genetic Algorithms. Van Nostrand Reinhold, New York (1991)
24. Deb, K.: Multi-Objective Optimization Using Evolutionary Algorithms. Wiley, West Sussex (2001)
25. Di Paolo, E.A., Noble, J., Bullock, S.: Simulation models as opaque thought experiments. Seventh International Conference on Artificial Life, pp. 497–506. MIT Press, Cambridge (2000)
26. Dray, A., Mazerolle, L., Perez1, P., Ritter, A.: Policing Australia's heroin drought: using an agent-based model to simulate alternative outcomes. J. Exp. Criminol. **4**, 267–287 (2008)
27. Eck, J., Weisburd, D.: Crime places in crime theory. In: Eck, J., Weisburd, D. (eds.) Crime and Place, pp. 1–33. Criminal Justice Press, Monsey (1995)
28. Eck, J.E., Liu, L.: Contrasting simulated and empirical experiments in crime prevention. J. Exp. Criminol. **4**(3), 195–213 (2008)
29. Egesdal, M., Fathauer, C., Louie, K., Neuman, J., Mohler, G., Lewis, E.: Statistical and Stochastic Modeling of Gang Rivalries in Los Angeles. SIAM Undergraduate Research Online (SIURO) 3. [Onine] (2010). http://www.siam.org/students/siuro/vol3/S01045.pdf. Accessed 16 June 2011
30. Eiben, A.E., Smith, J.E.: Introduction to Evolutionary Computing. Springer, Heidelberg (2003)
31. Efstratiadis, A., Koutsoyiannis, D.: On the practical use of multiobjective optimisation in hydrological model calibration, European Geosciences Union General Assembly 2009, Geophysical Research Abstracts, vol. 11. European Geosciences Union, Vienna (2009)
32. Epstein, J.M., Axtell, R.L.: Growing Artificial Societies: Social Science from the Bottom Up. MIT Press, Cambridge (1996)
33. Evans, A.J.: A sketchbook for ethics in agent-based modelling. Association of American Geographers (AAG) Annual Meeting, 23–26 February 2012, New York [online] (2012). http://www.geog.leeds.ac.uk/presentations/12-2/12-2.pptx

34. Expósito, R.R., Taboada, G.L., Ramos, S., Touriño, J., Doallo, R.: Performance analysis of HPC applications in the cloud. Future Gener. Comput. Syst. **29**(1), 218–229 (2013). doi:10.1016/j.future.2012.06.009
35. Farmer, J.D., Foley, D.: The economy needs agent-based modelling. Nature **460**, 685–686 (2009)
36. Grimm, V., Railsback, S.F.: Individual-Based Modeling and Ecology. Princeton University Press, Princeton (2005)
37. Grimm, V., Berger, U., Bastiansen, F., Eliassen, S., Ginot, V., Giske, J., Goss-Custard, J., Grand, T., Heinz, S., Huse, G., Huth, A., Jepsen, J.U., Jørgensen, C., Mooij, W.M., Müller, B., Pe'er, G., C., Piou, Railsback, S.F., Robbins, A.M., Robbins, M.M., Rossmanith, E., Rüger, N., Strand, E., Souissi, S., Stillman, R.A., Vabø, R., Visser, U., DeAngelis, D.L.: A standard protocol for describing individual-based and agent-based models. Ecol. Model. **198**, 115–126 (2006)
38. Groff, E.: Exploring The Geography Of Routine Activity Theory: A Spatio-Temporal Test Using Street Robbery. PhD thesis, University of Maryland (2006)
39. Groff, E.: Simulation for theory testing and experimentation: An example using routine activity theory and street robbery. Journal of Quantitative Criminology, **23**:75–103 (2007a)
40. Groff, E.: Situating simulation to model human spatio-temporal interactions: An example using crime events. Transactions in GIS, **11**(4):507–530 (2007b)
41. Groff, E., Mazerolle, L. (2008) Simulated experiments and their potential role in criminology and criminal justice. Journal of Experimental Criminology, **4**(3):187–193
42. Goldberg, D.: Genetic Algorithms: in Search, Optimisation and Machine Learning. Addison Wesley, Crawfordsville (1989)
43. Goldstein, N.C.: Brains versus brawn—comparative strategies for the calibration of a cellular automata-based urban growth model. In: Atkinson, P., Foody, G., Darby, S., Wu, F. (eds.) Geodynamics, pp. 249–272. CRC Press, Boca Raton (2004)
44. Hayslett-McCall, K. L., Qiu, F., Curtin, K. M., Chastain, B., Schubert, J., Carver, V.: The simulation of the journey to residential burglary. In: Artificial Crime Analysis Systems: Using Computer Simulations and Geographic Information Systems, chapter 14. IGI Global, Hershey (2008)
45. Heppenstall, A.J., Evans, A.J., Birkin, M.H.: Genetic algorithm optimisation of an agent-based model for simulating a retail market. Environ. Plan. B: Plan. Design **34**, 1051–1070 (2007)
46. Holland, J.: Adaption in Natural and Artificial Systems. MIT Press, Cambridge (1975)
47. Holland, J.: Genetic algorithms. Scientific American, July 1992, 66–72
48. Huddleston, S.H., Learmonth, G.P., Fox, J.: Changing Knives into Spoons. Proceedings of the 2008 IEEE Systems and Information Engineering Design Symposium, University of Virginia, Charlottesville, VA, USA, April 25 (2008) [Online]. http://www.sys.virginia.edu/sieds09/papers/0047_FPM2SimDM-02.pdf. Accessed 16 June 2011
49. Johnson, S.D., Bernasco, W., Bowers, K.J., Elffers, H., Ratcliffe, J., Rengert, G.F., Townsley, M.: Near repeats: a cross national assessment of residential burglary. J. Quant. Criminol. **23**(3), 201–219 (2007)
50. Johnson, S.D., Bowers, K.J.: Permeability and burglary risk: are Cul-de-sacs safer? J. Quant. Criminol. **26**(1), 89–111 (2009)
51. Knudsen, D.C., Fotheringham, A.S.: Matrix comparison, goodness-of-fit, and spatial interaction modeling. Int. Reg. Sci. Rev. **10**, 127–147 (1986)
52. Kongmuang, C.: Modelling crime: a spatial microsimulation approach. Ph.D. thesis, School of Geography, University of Leeds, Leeds (2006)
53. Li, X., Yang, Q.S., Liu, X.-P.: Genetic algorithms for determining the parameters of cellular automata in urban simulation. Sci. China Ser. D: Earth Sci. **50**(12), 1857–1866 (2007)
54. Liu, L., Wang, X., Eck, J., Liang, J.: Simulating crime events and crime patterns in a RA/CA models. In: Wang, F. (ed.) Geographic Information Systems and Crime Analysis, pp. 197–213. Idea Publishing, Reading (2005)
55. Lustick, I.S.: Defining Violence: a plausibility probe using agent-based modeling. Paper prepared for LiCEP, Princeton University, May 12–14, 2006 [Online]. http://www.prio.no/files/file48070_lustick_violdef_foroslo_v2.pdf. Accessed 16 June 2011

56. Malleson, N.: An agent-based model of burglary in Leeds. Master's thesis, University of Leeds, School of Computing, Leeds [Online] (2006). http://www.geog.leeds.ac.uk/fileadmin/downloads/school/people/postgrads/n.malleson/mscproj.pdf. Accessed 16 June 2011
57. Malleson, N.: Agent-based modelling of burglary. Ph.D. thesis, School of Geography, University of Leeds (2010)
58. Malleson, N., Evans, A.J., Jenkins, T.: An agent-based model of burglary. Environ. Plan. B: Plan. Des. **36**, 1103–1123 (2009)
59. Malleson, N., Heppenstall, A.J., Evans, A.J., See, L.M.: Evaluating an agent-based model of burglary. Working paper 10/1, School of Geography, University of Leeds, UK. January 2010 [Online]. http://www.geog.leeds.ac.uk/fileadmin/downloads/school/research/wpapers/10_1.pdf. Accessed 16 June 2011
60. Malleson, N., Heppenstall, A.J., See, L.M.: Crime reduction through simulation: an agent-based model of burglary. Comput. Environ. Urban Syst. **34**, 236–250 (2010b)
61. Malleson, N., See, L.M., Evans, A.J., Heppenstall, A.J.: Implementing comprehensive offender behaviour in a realistic agent-based model of burglary. Simul.: Trans. Soc. Model. Simul. Int. **88**(1), 50–71 (2010)
62. Malleson, N., Evans, A.J., Heppenstall, A.J., See, L.M.: Crime from the ground-up: agent-based models of burglary. Geographical Compass (Submitted)
63. Malleson, N., Evans, A., Heppenstall, A., See, L..: The Leeds Burglary Simulator. Informatica e diritto special issue: Law and Computational Social Science **1** 211–222 (2013)
64. Maros, I., Mitra, G.: Simplex algorithms. In: Beasley, J.E. (ed.) Advances in Linear and Integer Programming, pp. 1–46. Oxford Science, Oxford (1996)
65. Melo, A., Belchior, M., Furtado, V.: Analyzing police patrol routes by simulating the physical reorganization of agents. In: Sichman, J.S., Antunes, L. (eds.) MABS, Lecture Notes in Computer Science, vol. 3891, pp. 99–114. Springer, Heidelberg (2005)
66. Michalewicz, M.: Genetic Algorithms + Data Structures = Evolution Programs. Springer, Heidelberg (1992)
67. Michalewicz, Z., Janikow, C.: Genetic algorithms for numerical. Optim. Stat. Comput. **1**(2), 75–91 (1991)
68. Mitchell, M.: An Introduction to Genetic Algorithms. MIT Press, Cambridge (1998)
69. Moss, S., Edmonds, B.: Towards good social science. J. Artif. Soc. Soc. Simul. **8**(4), 13 (2005)
70. Omer, I.: How ethnicity influences residential distributions: an agent-based simulation. Environ. Plan. B: Plan. Des. **32**(5), 657–672 (2005)
71. Rengert, G.F., Wasilchick, J.: Suburban burglary: a time and a place for everything. Charles Thomas, Springfield (1985)
72. Schmidt, B.: The Modelling of Human Behaviour. SCS Publications, Erlangen (2000)
73. Schmidt, B.: How to give agents a personality. Proceedings of the 3rd Workshop on Agent-Based Simulation, April 7–9, Passau, Germany (2002)
74. Schelling, T.C.: Dynamic models of segregation. J. Math. Sociol. **1**, 143–186 (1971)
75. Shan, J., Alkheder, S., Wang, J.: Genetic algorithms for the calibration of cellular automata urban growth modeling. Photogram. Eng. Remote Sens. **74**(10), 1267–1277 (2008)
76. Shaw, C., McKay, H.: Juvenile Delinquency and Urban Areas. University of Chicago Press, Chicago (1942)
77. Stonedahl, F.J.: Genetic algorithms for the exploration of parameter spaces of agent-based models. Unpublished Ph.D. thesis, Northwestern University, Evanston (2011). http://forrest.stonedahl.com/thesis/forrest_stonedahl_thesis.pdf
78. Tesfatsion, L., Judd, K.L.: Handbook of Computational Economics: Agent-Based Computational Economics. North Holland, Amsterdam (2006)
79. Urban, C.: PECS: a reference model for the simulation of multi-agent systems. In: Suleiman, R., Troitzsch, K.G., Gilbert, N. (eds.) Tools and Techniques for Social Science Simulation, Chapter 6, pp. 83–114. Physica, Heidelberg (2000)
80. van Baal, P.: Computer Simulations of Criminal Deterrence: From Public Policy to Local Interaction to Individual Behaviour. Boom Juridische Uitgevers, Den Haag, The Netherlands (2004)

81. Vickers, D., Rees, P.: Creating the UK national statistics 2001 output area classification. J. Royal Stat. Soc. Ser. A **170**(2), 379–403 (2007)
82. Vrugt, J.A., Gupta, H.V., Bastidas, L.A., Bouten, W., Sorooshian, S.: Effective and efficient algorithm for multiobjective optimization of hydrologic models. Water Resour. Res. **39**(8), 1214 (2003). doi:10.1029/2002WR001746
83. Wang, X., Liu, L., Eck, J.: Crime simulation using gis and artificial intelligent agents. In: Liu, L., Eck, J. (eds.) Artificial Crime Analysis Systems: Using Computer Simulations and Geographic Information Systems, chapter 11. Information Science Reference, Hershey (2008)
84. Weisburd, D., Bushway, S., Lum, C., Ang, S.-M.: Trajectories of crime at places: a longitudinal study of street segments in the city of Seattle. Criminology **42**(1), 283–321 (2004)
85. Weisburd, D., Bruinsma, G.J.N., Bernasco, W.: Units of analysis in geographic criminology: historical development, critical issues, and open questions. In: Weisburd, D., Bernasco, W., Bruinsma, G.J.N. (eds.) Putting Crime in Its Place. Units of Analysis in Geographic Criminology (pp. 3–31). Springer, Heidelberg (2009)
86. Wiese, T.: Global Optimization Algorithms—Theory and Applications, 2nd edn. University of Kassel, Distributed Systems Group (2009). http://www.it-weise.de
87. Winoto, P.: A simulation of the market for offenses in multiagent systems: is zero crime rates attainable? In: Sichman, J.S., Bousquet, F., Davidsson, P., (eds.) MABS, Lecture Notes in Computer Science, vol. 2581, pp. 181–193. Springer, Heidelberg (2003)
88. Xiao, W.: A unified conceptual framework for geographical optimization using evolutionary algorithms. Ann. Assoc. Am. Geogr. **98**(4), 795–817 (2008)
89. Yapo, P.O., Gupta, H.V., Sorooshian, S.: Multi-objective global optimization for hydrological models. J. Hydrol. **204**, 83–97 (1998)

Printed in the United States
By Bookmasters